Soil Management for Sustainability

SOIL MANAGEMENT FOR SUSTAINABILITY

R. Lal and F. J. Pierce
Editors

SOIL AND WATER CONSERVATION SOCIETY

Published by the
Soil and Water Conservation Society
7515 Northeast Ankeny Road
Ankeny, Iowa 50021-9764
in cooperation with the
World Association of Soil and Water Conservation
and the
Soil Science Society of America

Manufactured in the United States of America

Library of Congress Catalog Card No. 90-25662

ISBN 0-935734-23-6

$15.00

Library of Congress Cataloging-in-Publication Data

Main entry under title:

Soil management for sustainability / R. Lal and F. J. Pierce, editors
Papers from a workshop in Edmonton, Alberta, Aug. 1989.
ISBN 0-935734-23-6
1. Soil management—Congresses. 2. Soil conservation—Congresses.
I. Lal, R. II. Pierce, F. J. (Fran J.) III. Soil and Water Conservation Society (U.S.) IV. World Association of Soil and Water Conservation. V. Soil Science Society of America
S590.2.S627 1991 90-25662
631.4'5—dc20 CIP

To Dr. William L. Larson
for his life-long dedication to soil
management and land stewardship

Contents

Foreword

This publication honors Dr. William E. Larson for his career-long devotion to land stewardship. I had the privilege of working with Dr. Larson from 1957, when I went to Ames, Iowa, with the U.S. Department of Agriculture-Agricultural Research Service, for most of my career. I was able to observe first-hand his dedication to and effectiveness in soil and water management research.

I remember in 1962, at the American Society of Agronomy annual meeting at Cornell University, I was rooming with Dr. Larson. I came in and he was rehearsing his presentation for the next day. His paper introduced his row-interrow concept. This was a classic for its time and is still widely cited today. This was the beginning of modelling. When computers came along, he had been waiting for them as the tool he needed.

Dr. Larson has contributed much to our understanding of soil interactions in so many areas. Perhaps his most important contribution has been his desire and ability to show how basic research in soil physics and soil chemistry can be applied to a better understanding of soil tillage, soil compaction, soil erosion, and myriad other aspects of soil management and productivity. He has made a contribution in many areas of crop and soil science besides tillage, compaction, and erosion. Among these are soil temperature, residue effects on plant growth and erosion, nitrification, aggregation, porosity and aeration, soil strength, soil structure, infiltration, soil mechanics, porous media in general, factors affecting root growth, sewage disposal on land, water quality, erosion-productivity relationships, productivity of dredged materials and soil-loss tolerance. For some years, Dr. Larson has been fascinated by data bases and their contribution to understanding of what is happening in the field.

Dr. Larson has received many honors throughout his years as a soil scientist. Among them are the most prestigious award of the Soil and Water Conservation Society—the Hugh Hammond Bennett Award—along with the Outstanding Service Award, the Fellow Award, and a President's Citation from this same Society; the Soil Science Award from the American Society of Agronomy; the Superior Service Award in Science and Technology from the U.S. Department of Agriculture; Honorary Doctor of Science Degree from the University of Nebraska; the Floyd Andre Award from Iowa State University; the Charles A. Black Award from the Council for Agricultural Science and Technology; as well as Fellowship in the Soil Science Society of America, American Society of Agronomy, and the American Association for the Advancement of Science. He has served as president of both the American Society of Agronomy and the Soil Science Society of America.

The World Association of Soil and Water Conservation chose to recognize Dr. Larson by dedicating to him the workshop on which this publication is based. These are all topics to which he has devoted much of his career and in which he has made a significant contribution. All are written by colleagues or former students who are authorities in their respective fields. The publication, besides highlighting Dr. Larson's brilliant career, updates a number of areas in soil management critical to maintaining a productive and sustainable soil resource base for future generations.

William C. Moldenhauer, Executive Secretary
World Association of Soil
and Water Conservation

Preface

The decade of the 1960s produced the term "green revolution." "Farming systems" emanated from the 1970s. The decades of the 1980s and 1990s will be remembered as the era of "sustainable agriculture." In fact, the term "sustainable" has become a buzzword. Donor agencies are seeking projects under this banner; policymakers are enthused about it; and scientists are procuring financial support using this terminiology to pursue researchable topics of interest to them.

Interest in sustainable agriculture may be overexaggerated; it could be another fad that will eventually fade away. However, the challenge of agricultural sustainability is neither new nor is it likely to fade away in the foreseeable future. Throughout civilization, the human race has been faced with the dilemma of meeting its current needs while preserving resources for generations to come. In fact, we are confronted by the same challenge more forcibly now than ever before. Rhetoric aside, our interests in sustainable agriculture arise out of mounting environmental hazards of modern agricultural practices, dwindling profit margins, rapidly degrading soil and water resources, and perpetual food deficits in some regions of developing countries, for example, sub-Saharan Africa.

Sustainable agriculture and the relevant environmental and socioeconomic concerns are global issues. Consequently, it has all the makings of an ideal concept that requires a holistic, integrated, interdisciplinary, or systems-oriented approach that can be talked about but not easily translated into practical research. While an attractive concept, it is difficult to identify an entry point into this complex problem of sustainable agriculture. One important entry point, however, must be the enhancement of soil quality and the improvement of its life-support processes. Dr. William E. Larson has been foremost among soil scientists who pioneered this concept of soil resource management for sustaining agricultural productivity while maintaining an acceptable standard of environmental quality. Not only did he envision this concept and attempt to articulate it long before it became a buzzword, but he also saw the need for an interdisciplinary approach to address this important problem.

Dr. Larson has been a founding member and an enthusiastic supporter of the World Association of Soil and Water Conservation. From its outset, WASWC has been concerned with sustainable management of soil and water resources. WASWC, therefore, deemed it fitting to organize a workshop, "Soil Management for Sustainability," in honor of Dr. Larson. The theme of the workshop and specific topics were carefully chosen to highlight research contributions made by Dr. Larson throughout a distinguished career that spanned four decades. Speakers were chosen on the basis of their professional competence and their collaboration and association with Dr. Larson.

This book consists of 14 manuscripts in all, 12 of which were presented at the workshop, held in Edmonton, Alberta, in August 1989. Introductory and concluding chapters were subsequently added to provide continuity and coherence. Topics discussed in the book can be conveniently divided into three parts: basic processes, management options, and policy issues and priorities. Important themes covered among basic processes include soil structure, soil compaction, and predicting soil erosion and its effects on crop productivity. Management options include conservation tillage, utilization of organic wastes, farming by soils, and agricultural sustainability. The final section deals with resource assessment policy and researchable priorities.

This book is a tribute to a colleague who has made outstanding contributions in soil management for sustainability. WASWC has been privileged to receive guidance and support from Dr. Larson and was pleased to be involved in organizing the workshop. Financial support for the workshop and for publication of this volume was received from the Soil Science Society of America—support we gratefully acknowledge. Several workshop participants were sponsored by the U.S. Department of Agriculture's Soil Conservation Service and the International Affairs Committee of the Soil and Water Conservation Society.

I very much appreciated the support provided by Dr. F. J. Pierce as well. Fran was involved in organizing the workshop, seeking financial support, contacting authors, and editing manuscripts. He extended his support with a sense of professional dedication and personal commitment.

My special thanks are due Dr. William C. Moldenhauer. His guidance and advice was valuable indeed.

It was a pleasure also for me to work with all of the authors involved in preparation of this volume. In spite of their busy schedule, they responded enthusiastically, prepared and revised papers promptly, and produced important manuscripts.

As usual, Fran Pierce and I received full cooperation and support from Max Schnepf, James Sanders, and Tim Kautza of the Soil and Water Conservation Society. It is always a pleasure to work with them and other members of their staff.

Rattan Lal, President
World Association of Soil
and Water Conservation

1

The Vanishing Resource

R. Lal and F. J. Pierce

Emphasis on agricultural sustainability arises out of increasing awareness about the finite nature of the world's arable land resources, the widespread problem of soil degradation, the rapidly deteriorating quality of the environment, the greenhouse effect, the excessive reliance on nonrenewable resources, and the need to preserve soil resources for long-term use rather than exploitation for short-term gain. Agricultural sustainability issues also relate to dwindling profit margins in large-scale, commercial farming in the developed countries and to the vicious cycle of low yield, soil degradation, malnutrition, and famine in several parts of the developing world. Furthermore, most issues related to sustainable management of natural resources are global; they cut across political and national boundaries and require a coordinated, long-term effort to understand the cause-effect relationship and to develop meaningful solutions.

Major Issues

Despite the ever-increasing demand for prime agricultural land, the earth's land resources are not only finite but also nonrenewable relative to the present civilization. Currently, about 1.5 billion hectares are now cropped worldwide. Much of the potentially arable land is marginal for agricultural use because it is either inaccessible, too steep, too shallow, or located in regions with too little or too much water. Essential inputs for sustained production often are not available as well.

The United States has about 170 million hectares in cropland, another 62 million hectares with medium to high potential for conversion to cropland, and an additional 138 million hectares with low conversion potential (*18*). In 1982 the agricultural production consumed could have been produced on 80 percent of the 155 million hectares harvested (*18*). Over-production in the United States, combined with what appears to be an adequate natural resource base, diminishes the urgency of the global situation in developed and developing countries alike. Brown and associates (*3*) warn, however, that nations, such as the United States, are living under the illusion of progress.

Considering the total world population of about 5 billion in 1986, the per capita arable land area amounts to a meager 0.3 hectare. The per capita arable land area will progressively decline to 0.23 hectare by the year 2000, to 0.15 hectare by 2050, and to about 0.14 hectare by the time world population stabilizes in the year 2100. The change in per capita grain production declined from +13 percent per decade in the 1950s to −2 percent per decade in the 1980s;

for the 1990s it is projected to be −7 percent (*3*). It is estimated that the minimum per capita arable land needed for an adequate diet is 0.5 hectare under a modest level of input. Will agricultural technologies enable production of at least basic necessities from per capita arable land of 0.25 or 0.15 hectare? What is the view from developed nations, such as the United States, which has per capita arable land of 0.68 hectare?

The problem of land-to-people ratio is further compounded by the serious and global problem of soil degradation. Arable land resources are continuously being depleted by several processes of soil degradation: erosion by wind and water, anaerobiosis, salinization/alkalization, compaction and hard-setting, depletion of soil organic matter, and nutrient imbalance. It is often argued that, aside from the 1.5 billion hectare of currently cropped land, another 2.0 billion hectare of once biologically productive land have been rendered unproductive through irreversible degradation (*17*). Some surveys have estimated that soil degradation, of one type or another, affects about one-third of Earth's land surface.

Mismanagement, neglect, and exploitation can ruin the fragile resource and become a threat to human survival. In fact, archaeological evidence has shown that soil degradation was responsible for extinction of the Harappan civilization in western India, Mesopotamia in western Asia, and the Mayan culture in Central America (*11*). Furthermore, the current rate of soil degradation is estimated at five to seven million hectares per year, and the annual rate may increase to 10 million hectares by the year 2000 (*6*). Brown and colleagues (*3*) estimate that the world could be losing 14 million tons of grain output because of environmental degradation, primarily soil erosion.

Accelerated erosion is a major factor responsible for soil degradation (*8, 13*). Accelerated soil erosion has irreversibly destroyed an estimated 430 million hectares in different countries. This is about 30 percent of the presently cultivated land area. Worldwide, natural erosion is estimated to total some 9.9 billion tons of soil a year, but human-induced, accelerated erosion is more than 2.5 times higher—26 billion tons per year (*2, 19*). Some of the ecologically sensitive regions include Ethiopia, Haiti, the Andean region, countries in the African Sahel, and the Himalayan-Tibetan ecosystem. In addition, there is the issue of tropical rainforest depletion—at the rate of about 11 million hectares per year. Most of this newly cleared land is prone to severe erosion (*9*). Erosion is a major limitation to the use cf cropland in the United States, where 52 percent of cropland is classified as erodible ("e" subclasses) in the U.S. Department of Agriculture's land capability classification system and 31 million hectares erode at rates that result in a five percent or more loss of their productivity within 100 years (*18*).

In addition to direct economic consequences related to reduced productivity, soil degradation has severe environmental consequences. However, these consequences have not been given the emphasis they deserve and raise questions about the environmental and ethical soundness of agricultural production practices. For example, the recovery of nitrogenus fertilizers is only 50 to 60 percent and often as low as 15 percent. The unrecovered fertilizers and pesticides are often leached to groundwater supplies or transported in runoff to surface water. The annual discharge of pollutants to waterways from agricultural land in the United States is about 1,100 million tons of suspended solids, 500 million tons of dissolved soils, 1.2 million tons of total phosphorus (P) and 4.7 million tons of total nitrogen (N) (*5*). Clark and associates (*4*) concluded that the off-farm impacts of soil erosion in the United States exceeded $6 billion annually and far exceeded the on-farm effects due, for example, to loss in productivity. Over the last two decades, at least 46 pesticides have been detected in groundwater in 26 states in the United States, and nitrate contamination is far more common (*12*). Concern about agrichemical contamination of groundwater and related issues, such as food safety and surface water quality, is high in developed countries, such as the United States, and the stage is set for global concern.

The use of pesticides and fertilizer is rapidly increasing in Third World countries. It is estimated that about 25 percent of the world's pesticide market is in developing countries, and pesticide use is expected to more than double by the year 2000 (*10*). Despite the rapidly increasing demand, there is little or no attempt to develop effective policies and legislation to regulate the use of these compounds. Consequently, there is great danger of contamination of surface water and groundwater by water-soluble pesticides and fertilizers. Contaminated supplies of drinking water in the Third World can be a major health hazard to a vast majority of the population.

Excessive energy use in agriculture is also a major economic and environmental concern. Modern agriculture is an energy-intensive activity. The energy input per hectare of corn production in the United States is estimated at 7 million K cals (*7, 13*). Ritchie (*14*) estimated that the percentage of total energy used in production is 30 to 38 for fertilizer, 15 to 17 for crop drying, 10 to 15 for field operations, 7 to 13 for supplemental irrigation, 2 to 5 for pesticides, and 15 to 20 for other operations. Most of this energy used in agriculture is derived from nonrenewable fossil fuels. Most energy-intensive systems, though highly productive, are least efficient thermodynamically.

Although scientists have recognized the dangers of global warming caused by the burning of fossil fuel, the impact of the emission of carbon dioxide and other greenhouse gases (methane, carbon monoxide, nitrogen oxide) from soil-related processes has hitherto been ignored. Predominant soils in the tropical regions (Oxisols, Ultisols, Alfisols, Aridisols, Entisols, Inceptisols and Mollisols) occupy about 4.2 billion hectares, with a weighted mean average organic carbon content of 1.45 percent in the top layers. If 1 percent of this organic carbon (1.45 percent) is mineralized each year due to anthropological perturbations, 128 billion tons of carbon will be emitted into the atmosphere. This compares with an estimated 325 million tons from fossil fuels and 1,659 million tons from deforestation (*3*). On the other hand, the same amount of carbon can be fixed from the atmosphere in the soil by adoption of sustainable soil and water management systems. Soil is, therefore, a potential source or sink for carbon depending on the management systems adopted. There is an absolute lack of knowledge about the contribution of soil-related processes to emission of greenhouse gases under different systems of soil and crop management. Contributions to atmospheric carbon dioxide by soil organic matter is anyone's guess.

Some solutions

In spite of uncertainty about the estimates cited above, there is need for immediate action by the international community to do something about these major problems. Sustainable soil and water management systems must be developed, not only to curtail further degradation, but also to restore the productivity of land already degraded. Mechanized, fossil fuel-based agriculture is faced with serious questions pertaining to economic and environmental issues. Thus, there is an increasing awareness of the need for developing and adapting agricultural systems based on judicious off-farm inputs and proper management of soil and water resources. Economic profitability of these new systems must be related to soil productivity and environmental quality. Unless significant advances are made in understanding the long-term impact of these new farming strategies, the potential for damage to soil and environment by agricultural ecosystems can be alarming, especially within or adjacent to sensitive and fragile ecosystems in the tropics and subtropics. Continuous, in tensive farming activities in fragile ecosystems, necessitated by high demographic pressure, accelerate soil erosion and may degrade a soil's life support processes beyond the point of no return.

The scientific community must develop agricultural technologies to (a) reduce input while maximizing economic returns, (b) decrease soil degradation, (c) minimize risks of pollution of natural waters and environments, (d) restore productivity of degraded land, and (e) maintain productive capacity of existing land by preserving a soil's life-support processes. As Robertson and Laverty (*15*) suggest, failure to control land and environmental degradation is also rooted in the economic, social, and political spheres; any approach to resolving degradation problems must extend well beyond the usually recognized boundary of the agriculture-food system.

Need For Sustainable Agriculture

Sustainable agriculture, currently receiving an enthusiastic response in the scientific community and from policymakers, has become a buzz word for the 1980s and 1990s. It is a philosophy aimed at establishing an ecological approach to resource management. The objective is to balance the inherent soil resource and crop requirements in innovative soil and crop management systems. The emphasis is not on maximizing production but on optimizing resource use and sustaining productivity over a long period.

There are several definitions of sustainable agriculture. Just as there is no universal blueprint of technological options for sustaining agricultural production for all soils and ecological regions, similarly there is no single definition that is widely accepted or applicable. A commonly used definition is that proposed by the Technical Advisory Committee to the Consultative Group on International Agricultural Research. According to this definition, "sustainable agriculture should involve the successful management of resources to satisfy changing human needs while maintaining or enhancing the quality of the environment and conserving natural resources" (*16*).

Sustainable agriculture is not necessarily synonymous with low-input agriculture. Reducing input is not an option for many Third World countries where no inputs or low inputs are now used for crop production. A clear distinction, therefore, must be made in creating technological options for sustainable agricultural production in developing versus developed countries. Furthermore, lowering agricultural resource inputs below certain levels in developed countries will result in lower yields. Increased use of land is a likely consequence of lower yields if production levels are to be maintained. If U.S. farmers were still harvesting the same annual yield per hectare in the 1980s as they did in the 1930s, an additional 170 million hectares would be required to produce the same volume of agricultural crops (*1*). The new land to be brought under cultivation could be marginal, prone to erosion and other degradative processes, and not suitable for agriculture.

Low-input technologies being recommended are not necessarily free from risks of environmental pollution. Frequent and heavy applications of organic manures, plowing under of leguminous crops as green manure, and runoff from feedlots and grazed pastures are equally as hazardous to the environment as are synthetic chemicals. Careful consideration must be given to agricultural practices that help to reduce the contamination of ground and surface water (*12*).

There also are major socioeconomic considerations. A major consideration in the design of sustainable systems is the impact on the labor force. Energy and agrichemicals have substantially reduced the need for on-farm labor. What are the impacts of sustainable agriculture on the rural and urban economics if there is a shift in labor and energy and material flows in agriculture? How does sustainable agriculture affect the quantity and quality of food and its global distribution? We must recognize that the implications of shifts to sustainability transcend the normal boundaries of agricultural systems.

Yet, we also realize the need to protect our soil resources. To do that, continued effort is needed to determine the appropriate technology that will reduce the risks of soil and environmental degradation and produce positive economic returns. In view of the severe problems of soil degradation and potential for environmental problems as a result of our fossil fuel-based agriculture, there is a need to assess the long-term impact of conventional versus low-input, regenerative, or alternative agricultural systems. The major objectives are to adjust input-to-yield ratios to improve efficiency, preserve soil productivity, and minimize environmental pollution. All three issues are relative, have subjective values that vary among agroecological regions, and require long-term research.

REFERENCES CITED

1. Barrons, K. C. 1988. *Body building of soils.* Journal Science of Food and Agriculture. 6: 22-26.
2. Brown, L. R. 1984. *State of the world, 1984.* Worldwatch Institute, W. W. Norton & Company, New York, New York.
3. Brown, L. R., A. Durning, C. Flavin, H. French, J. Jacobson, M. Lowe, S. Postel, M. Renner, L. Starke, and J. Young. 1990. *State of the world, 1990.* Worldwatch Institute and W. W. Norton & Company, New York, New York.
4. Clark, E. H., II, J. A. Haverkamp, and W. Chapman. 1985. *Eroding soils: The off-farm impacts.* The Conservation Foundation, Washington, D.C.
5. Duttweiler, D. W., and H. P. Nicholson 1983. *Environmental problems and issues of agricultural non-point pollution.* In F. W. Schaller and G. W. Bailey [editors] *Agricultural Management and Water Quality.* Iowa State University Press, Ames. pp. 3-16.
6. Food and Agriculture Organization. 1983. *Guidelines for control of soil degradation.* Rome, Italy.
7. Goswami, D. Y. [editor] 1986. *Alternative energy in agriulture.* CRC Press, Inc. Boca Raton, Florida. 226 pp.
8. Kovda, V. A. 1980. *Land aridization and drought control.* Westview Press, Boulder, Colorado.
9. Lal, R. 1981. *Deforestation and hydrological problems.* In R. Lal and E. W. Russell [editors] *Tropical Agricultural Hydrology.* J. Wiley and Sons, London, England. pp. 131-140.
10. Moore, T. W. 1986. *The changing environment.* Springer Verlaag, New York, New York. 239 pp.
11. Olson, G. W. 1981. *Archaeology: Lessons on future soil use.* Journal Soil and Water Conservation 36: 261-264.
12. Office of Technology Assessment, U.S. Congress. 1990. *Beneath the bottom line. Agricultural approaches to reduce agrichemical contamination of groundwater.* Washington, D.C.
13. Pimentel, D., and C. W. Hall. 1984. *Food and energy resources.* Academic Press, New York, New York. 268 pp.
14. Ritchie, T. W. 1983. *Source base for farm energy alternatives.* McGraw Hill Base Co., New York, New York. pp. 8-1 to 8-6.
15. Robertson, J. A., and D. H. Laverty. 1988. *Land degradation—definition and overview. In Land Degradation and Conservation Tillage.* Canadian Soil Science Society, Location. pp. 1-11.
16. Technical Advisory Committee to the Consultative Group on International Agricultural Research. 1988. *Sustainable agricultural production: Implications for international agricultural research.* Report of the Technical Advisory Committee. CGIAR, Washington, D.C.
17. United Nations Environment Program. 1986. *Farming systems principles for improved food production and the control of soil degradation in arid and semi arid tropics.* ICRISAT, Hyderabad, India.
18. U.S. Department of Agriculture. 1989. *The second RCA appraisal. Soil, water, and related resources on nonfederal land in the United States. Analysis of conditions and trends.* Washington, D.C.
19. Walling, D. E. 1984. *The sediment yield of African rivers.* Publ. 144. International Association of Hydrologic Sciences, Wallingford, United Kingdom. pp. 265-283.

2

Soil Structure: Processes and Management

W. M. Edwards

Soil structure has been defined as the size and arrangement of particles and pores in soils. The durability of the structural units helps define the permanence of the structural framework within which many chemical, biological, and physical processes take place. Therefore, researchers attempt to understand and manage soil structure to facilitate soil use, especially with respect to crop production and the movement of air, water, and chemicals.

Management and Soil Structure

As Lal and associates (*27*) showed, the definition of soil structure and the interactions among many structure-related processes and properties vary widely in space and time and are of interest to researchers in many science fields. Machines and tillage implements used to prepare fields for crop production affect soil structure, as do the growth and decay of plants. Organic matter, in general, has a positive effect on aggregate strength, and under various conditions, organo-mineral complexes, polysaccharides, micorrhizae, and root/hyphae play important roles. Cropping sequences (rotations), crop management practices, fertility, drainage, irrigation, and the quality of water that interacts with soil aggregates affect their size and strength. Soil amendments, notably sludges and organic polymers, influence structure and have been applied to soils to effect structural changes. For many years, wetting/drying and freezing/thawing effects on soil structure and aggregate stability have been investigated. Stress-strain relations for individual aggregates, prepared soil blocks, and undisturbed soil samples can be measured with laboratory equipment, whereas individual pores and porosity have been characterized using thin sections or photo-image analysis of embedded samples.

Tisdale and Oades (*46*) reviewed soil-aggregate stability research and concluded that water-stable aggregates from 1 to 10 millimeters in diameter were necessary to create good structure for crop growth. The aggregates need to contain enough large pores to remain aerobic and enough small pores to hold plant-available water. Pore space between aggregates should allow for rapid infiltration. Aggregates smaller than 0.25 millimeters in diameter are held together by organo-mineral complexes and polysaccharides. With these agents, binding is relatively permanent and not influenced by organic-matter content changes related to tillage. Larger aggregates are mainly held together by roots and hyphae. Stability of these aggregates declines with organic matter content, as roots die and are not replaced.

For most mineral soils, structural stability decreases when farming is performed in a way that reduces organic matter content (*18*). Although Kemper and associates (*25*) reported conditions under which organic matter slows the recovery of cohesion in dispersed soils, organic matter loss associated with intensive cultivation has led not only to weaker structure but also to reduced yields (*19, 30*). Grass (sod) crops improve aggregation better than any other crop (*2*). Conversely, bare fallow is the most deleterious management treatment with respect to soil structure (*32*).

It has been widely accepted that soil fauna, especially earthworms, generally enhance soil structure (*9, 11, 24, 29*). However, Shaw and Pawluk (*40*) reported that earthworms can positively or negatively influence soil structure, depending on the species of earthworms and/or the nature of the soil. The stability of fresh earthworm casts, which eventually become components of soil structure, can vary tremendously, depending upon aging conditions and the earthworm's diet (*41*). Further evidence of the complexity of earthworm effects was presented by Hamilton and associates (*20*), who studied earthworm interactions in soil microcosms that had received sludge amendments, and by Hartenstein (*21*), who reviewed and interpreted global earthworm biotechnology.

Interactions among management and tillage practices involved in crop establishment and harvest influence soil structure and water movement in soils (*35*). Compaction from farm machinery usually reduces macroporosity. In many soils, the resulting lower permeability to water, air, and rooting reduces crop yields. In Ontario, Canada, compaction generated at harvest in the fall did not hurt structure or yield, but these properties were affected during spring compaction (*44*). The effects of heavy equipment and tillage operations on soil structure at various times of the year are currently being studied under a wide range of soil and environmental conditions in Minnesota (*47*), North Carolina (*7*), and Ohio (*15*).

Seasonal processes, especially soil freezing and thawing, have been studied in the northwestern United States and thoroughly reviewed by Bullock and associates (*5*). Interactions among the availability of soil water and freezing/thawing cycles strongly determine the resultant soil-structural conditions (*28*).

To some extent, soil physical properties, including structure, can be managed with operations that do not necessarily include physical disturbances, such as those induced by mechanical tillage operations. Soil additives can affect structure. Cationic, anionic, and nonionic polymers have been applied to soils to influence soil strength and water movement through the soil (*38, 45*). Piccolo and Mbagwu (*34*) examined the interaction between humic acids, which usually have a positive effect on soil structure, and both anionic and nonionic surfactants. At high treatment rates on some soils, humic acid limited the negative effects of the anionic surfactant and was additive to the slight, positive effect of the nonionic surfactant. These researchers suggest that adding humic acids to surfactants may have practical significance for soil management systems in which surfactants are used as components of herbicide formulations.

Other organic amendments are currently being studied to determine their effects on soil structure, including papermill sludge in Maine (*48*), sewage sludge in Texas (6), and coal ash in Wisconsin (*33*). In Minnesota, many aspects of sludge application to cropland, including its effects on soil structure, were evaluated (*10*). The stability of soil aggregates as affected by electrolytic concentration of soil water is

Table 1. Annual precipitation and runoff for continuous corn watersheds, no-till and conventional tillage, 1979-1988.

		Runoff	
Year	*Precipitation (mm)*	*No-till (mm)*	*Conventional (mm)*
1979	1124	3.81	140.2
1980	1175	4.90	312.8
1981	1057	0.14	142.2
1982	889	0.0	113.2
1983	1027	0.0	*
1984	909	2.31	*
1985	929	0.01	*
1986	966	9.23	*
1987	841	0.15	*
1988	854	0.03	*

*Not conventional-tillage corn.

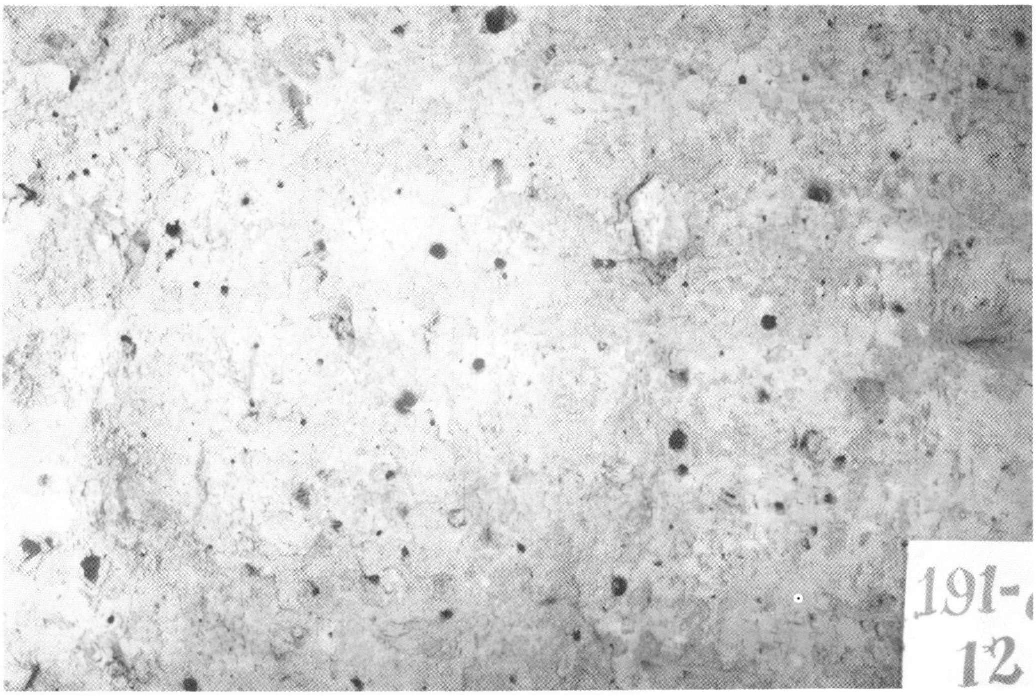

Figure 1. Macropores at the 30-cm depth under long-term no-till corn.

currently being studied in California (*1*) and Texas (*31*).

Water Movement through Structured Soils

In a practical sense, soil management operations on cropland are often aimed at producing a direct, desired agronomic response. However, as described above, scientists often measure the effects of management on physical or chemical properties and processes in the soil, which, in turn, hopefully can explain why and under what conditions the desired agronomic responses occur. Although the land manager may or may not be aware of the intermediate cause-and-effect relationships between the applied management practice and the agronomic response, researchers frequently try to relate the direct effects in functional models.

The no-till practice for growing corn (*Zea mays L.*) on the sloping fields of the eastern United States is a good example of a management practice that brings about the desired goal for the land manager, while providing the soil scientist with documented management-structure cause-and-effect relationships. At the North Appalachian Experimental Watershed in Coshocton, Ohio, watersheds where corn is grown under the no-till practice produce much less runoff than do other watersheds where conventional tillage practices are used (*12*). The no-till data in table 1 are derived from a watershed that has been in continuous no-till corn since 1964. The conventionally tilled watershed was plowed each year from 1978 to 1982, but it was necessary to return to meadow production in 1983 to help control the severe erosion that accompanied the high runoff. Although the two watersheds were not hydrologically identical, it appears that the direct runoff control objectives of the no-till management practice were met (4-year runoff totals were 708 millimeters for conventional tillage versus 9 millimeters for no-till). Pronounced differences in surface runoff accompany some less obvious subsurface effects of the management practice on soil properties.

Examination of the soil in the no-till watersheds shows many, nearly vertical earthworm burrows that appear as open, continuous channels through the upper and middle soil horizons (Figure 1). Most of the burrows made by *Lumbricus terrestris* L. are from 0.5 to 1 meter deep and greater than 5 millimeters in diameter. The no-till practice creates a near-surface environment that favors the *L. terrestris* population and the persistence of their burrows. Cultivation in the conventionally tilled watersheds destroys the burrows in the plow layer and reduces the amount of residue that remains on the surface, which these earthworms need as a food source.

Figure 1 shows a horizontal, 30-by-45 centimeter surface at the 30-centimeter depth in the long-term no-till watershed that supplied the runoff data for table 1. The larger, dark, nearly round areas in figure 1 are *L. terrestris* burrows. The number of these burrows greater than 5 millimeters in diameter have been estimated at 160 per square meter in this watershed (*12*).

Another consequence of the no-till manage ment practice on these soils involves a change in the distribution of density versus depth (Figure 2). With conventional tillage, the topsoil is loosened every year during pre-plant tillage, making it susceptible to crusting and near-surface compaction by summer storms. In contrast, the lower part of the plow layer in the no-till cornfields is compacted by machinery traffic and not loosened each year by tillage. Without tillage, crop residues accumulate at and near the surface, decreasing bulk density, reducing crust development, and helping to maintain high infiltration rates.

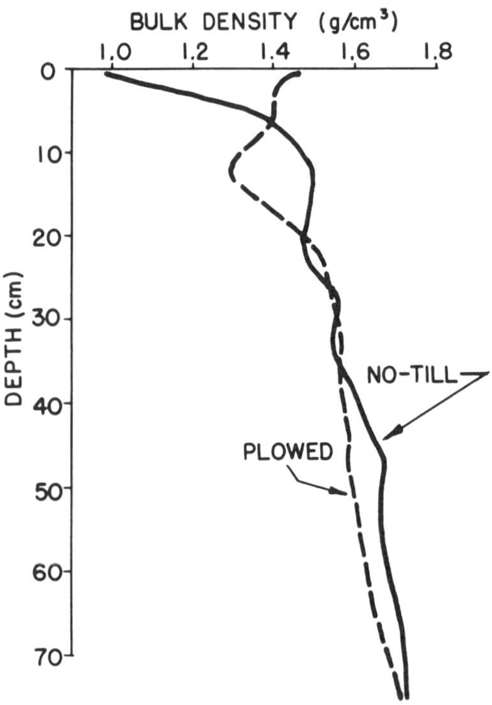

Figure 2. Bulk density versus depth in a Westmoreland silt loam farmed with continuous no-till and conventional tillage corn grown in rotation.

The measured macropore and density conditions created by the no-till management practice are but two, easily observed changes in soil-structure-related properties. The lack of disruptive tillage and the resulting persistent residue cover at the no-till surface create a much more favorable near-surface environment for many soil organisms than that of the conventionally tilled topsoil (*29*). Soil ecologists have documented these differences, many of which influence water movement near the soil surface (*9*).

If topsoils were homogeneous and isotropic, modeling water flow through them would be relatively easy. However, because a single 0.3-millimeter-diameter continuous macropore can affect saturated conductivity more than the entire matrix of a 30-centimeter- diameter undisturbed soil core (*43*), macropores, such as those shown in figure 1, must be considered potentially important flow pathways through the soil, at least under some conditions.

Because the soil is rarely homogeneous, infiltrating water moves deeper faster in some areas than in others (*39*). In general, the enhanced infiltration areas can be called preferential flow paths. They can result from spatial differences in conductivity, caused by variation in unidentified properties too subtle to be visually noted or conventionally measured (*23, 37*). Preferential flow can also result from flow instabilities at the wetting front (*22*). It has been noted that, even in very uniform sands, a small percentage of the surfce area may carry most of the infiltrating water (*26*).

In many cases, however, the reason for preferential flow is obvious. Shrinkage cracks in

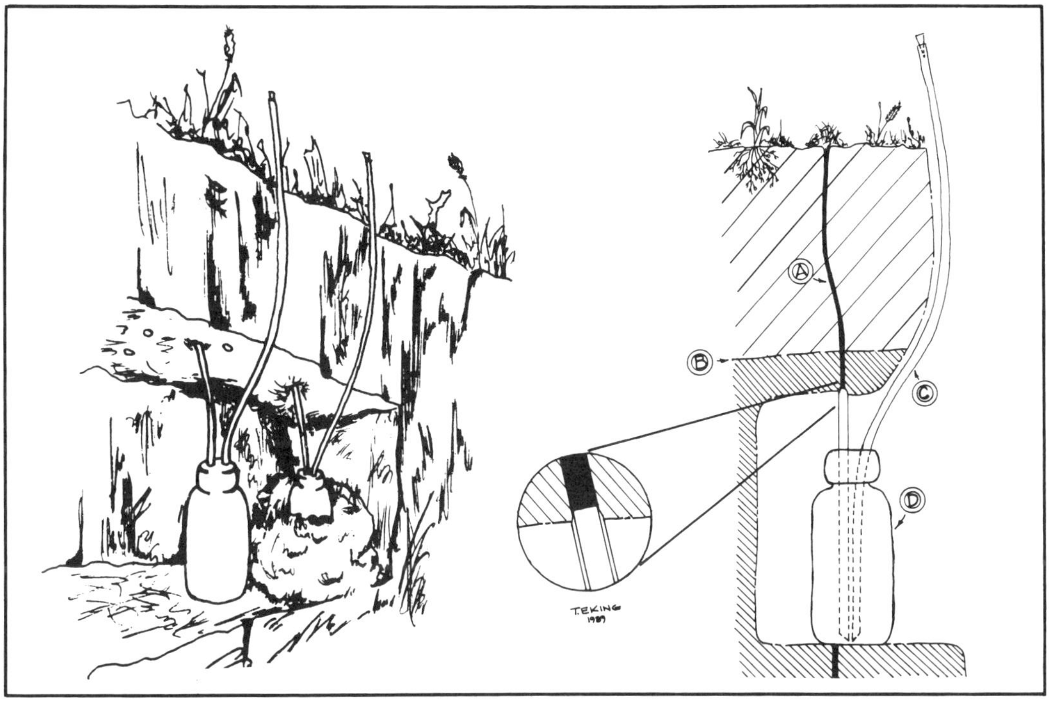

Figure 3. Earthworm burrow flow sampler. Schematic of installation before backfilling: (A) *L. terrestris* burrow, (B) old plowsole, (C) sampler access tube, and (D) sample bottle (*13*).

dry montmorillonitic soils act as obvious preferential flow paths during sudden, high-intensity storms. In forest soils, water has been observed flowing in voids left by decaying roots (*8, 16*). Cracks in rocks beneath soil profiles, as well as those between columnar structural units within the solum, clearly are preferential flow paths (*36*). Beven and Germann (*3*) described various kinds of preferential flow paths, including macropores and reviewed literature documenting water flow. They also discussed the importance of the flow characteristics of the soil as it influences movement into and out of the macropores.

The importance of *L. terrestris* burrows on the movement of water and chemicals into and through the no-till soils has been investigated at Coshocton, Ohio. Colored water, sprinkled on the no-till surface, lined earthworm burrows all the way to the bedrock contact, even when applied at low rainfall rates (*17*).

To investigate conditions under which natural rainfall events caused infiltration in the earthworm burrows, samplers (Figure 3) were installed below the 30-centimeter depth in 50 *L. terrestris* burrows in no-till corn fields (*13*). The samplers were installed in such a way that infiltrating flow conditions in the burrows and the surrounding soil above the sampled depth were not disturbed. Twelve growing-season storms during 1987 caused flow in the instrumented burrows, which was measured and analyzed to define the movement of water and nitrogen (Table 2). Worm-hole flow increased with storm size, but even small storms (3.6 millimeters of rain in 18 minutes) caused measured flow in some of the instrumented burrows. An average of 4 percent of the total rainfall from 12 summer storms, but as much as 10 percent of the rain falling in short, high-intensity storms, could be accounted for in the worm-hole flow. Edwards and associates (*13*) provide details of the worm-hole flow samplers, their installation, and the storm conditions that caused flow in the instrumented burrows.

Potential groundwater contamination by preferential flow in macropores spurs investigations into the quality of the rapidly infiltrating water. Bouma and associates (*4*) called the process "short circuiting," implying that the preferential flow could by-pass the filtering capacity of the fine-pore matrix. Particularly in continuous no-till corn fields, in which annual inputs of fertilizers and pesticides are broadcast on the surface, macropore flow may be suspected of transporting important amounts of chemicals through upper soil horizons. However, research has not yet established a generally applicable link between flow in earthworm burrows and groundwater contamination.

Possibly because of water movement in earthworm burrows, the effect of surface lime applications could be measured 50 centimeters below the old plow depth in a no-till field, whereas pH did not rise below the plow sole in adjacent fields that were periodically plowed (Figure 4) (*14*). The two fields contain the same soil, Westmoreland silt loam (fine-loamy, mixed, mesic Ultic Hapludalf), are within 200 meters of one another, and had both been farmed in a corn-wheat (*Triticum aestivum* L.)-meadow-meadow rotation prior to 1964 when the no-till field was shifted to continuous corn. In the next 20 years, the no-till field received four applications of lime at the rate of 4 megograms per hectare. The conventionally tilled field received three similar lime treatments in that period, after which the comparisons of pH (Figure 4) and of bulk density (Figure 2) were made. The depth of the pH effect in the no-till field approximates that of the *L. terrestris* burrows.

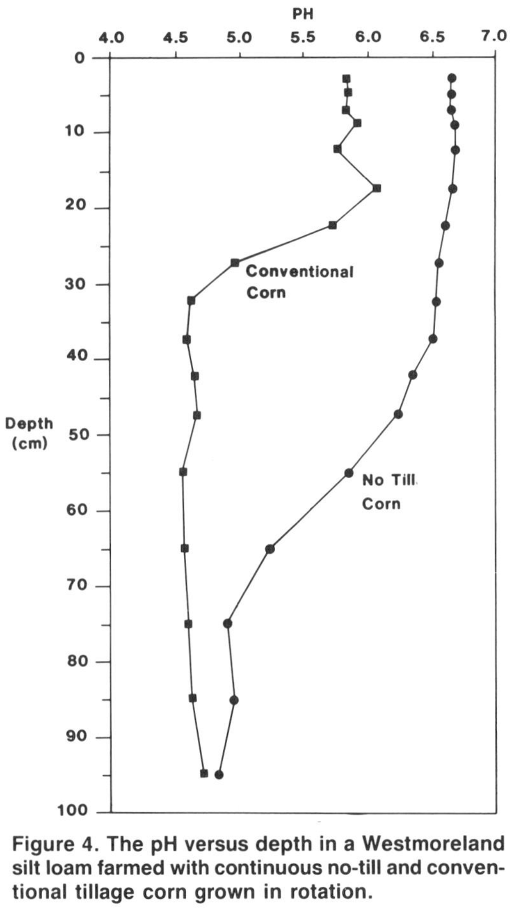

Figure 4. The pH versus depth in a Westmoreland silt loam farmed with continuous no-till and conventional tillage corn grown in rotation.

In a summer of relatively normal rainfall, less than 1 kilogram per hectare of nitrate nitrogen moved down to the 30-centimeter sampling depth in instrumented *L. terrestris* burrows (Table 2). Nitrate concentrations of sampled worm-hole flow during the next year, following a drought, were as much as 10 times greater than during the previous year. Ongoing work in the laboratory, using a rainfall simulator and blocks of undisturbed soil from no-till corn fields, indicates that the first storm following the surface application of chemicals moves higher concentrations of chemicals in the worm holes than do later events (*42*) and that the lining of the *L. terrestris* burrow itself may influence the quality of the infiltrating water.

Summary

Soil structure research is broadly focused over many disciplines involving interests and activities that either directly or indirectly influence the arrangement of particles and pores in the soil. The lack of annual tillage, as provided under continuous no-till management, encourages the development and persistence of macropores and other preferential flow paths that are potentially important for the downward movement of water and chemicals. As research further defines the effects of management on soil structure, the capability and the responsibility for managing the soil must be evaluated from ever-widening viewpoints.

REFERENCES CITED

1. Abu-Sharar, T. M., F. T. Bingham, and J. D. Rhoades. 1987. *Stability of soil aggregates as affected by electrolyte*

Table 2. Storm and sample characteristics of flow collected in 50 *L. terrestris* burrows in 12 storms, June-October 1987 (*13*).

Storm Date 1987	*Rainfall Amount (mm)*	*Rainfall Duration (hr)*	*Contributing Holes (no.)*	*Hole Flow Measured* (ml)*	*Hole Flow Measured* (ml/hole)*	*Hole Flow Calculated (L/ha)*	*Hole Flow Calculated (% of rain)*	*No³-N Concentration (mg/L)*	*No³-N Flux (g/ha)*
6/12	27.7	8.3	14	106	2.1	3,400	1.2	9.1	31
6/19	46.5	4.5	20	425	8.5	13,600	2.9	4.7	64
6/29	3.6	0.3	7	115	2.3	3,700	10.3	0.5	2
7/1	52.1	8.0	24	575	11.5	18,400	3.5	8.0	147
7/5	7.6	1.0	6	125	2.5	4,000	5.3	0.0	0
7/14	14.0	1.3	20	167	3.3	5,300	3.8	9.3	49
8/4	15.7	2.7	9	84	1.7	2,700	1.7	4.6	12
8/21	33.5	1.2	15	350	7.0	11,200	3.3	12.6	141
8/26	19.8	3.5	15	263	5.3	8,500	4.3	8.7	74
8/27	18.5	5.5	12	159	3.2	5,100	2.8	3.5	18
9/12	29.5	2.0	16	335	6.7	10,700	3.6	9.0	96
10/6	20.1	3.0	14	233	4.7	7,500	3.7	10.2	77
Total	288.6	41.3	172	2,937	58.8	94,100			711
Average	24.6	3.4	14.3	244.8	4.9	7,800	3.9		

*Total and average hole flow both measured for 50 holes.

concentration. Soil Science Society of America Journal 51(2): 309-314.

2. Allison, F. E. 1968. *Soil aggregation—some facts and fallacies as seen by a microbiologist.* Soil Science 106(2): 136-143.
3. Beven, K., and P. Germann. 1982. *Macropores and water flow in soils.* Water Resources Research 18(6): 1,311-1,325.
4. Bouma, J., L. W. Dekker, and C. J. Muilwijk. 1981. *A field method for measuring short-circuiting in clay soils.* Journal of Hydrology 52: 347-354.
5. Bullock, M. S., W. D. Kemper, and S. D. Nelson. 1988. *Soil cohesion as affected by freezing, water content, time and tillage.* Soil Science Society of America Journal 52(3): 770-776.
6. Cocke, C. L., and K. W. Brown. 1986. *Effect of sewage sludge on the physical properties of lignite overburden.* Reclamation Revegetation Research 6: 83-93.
7. Denton, H. P., and D. K. Cassel. 1989. *Conservation tillage and soil physical properties.* In M. G. Cook and W. M. Lewis [editors] *Conservation Tillage for Crop Production in North Carolina.* Bulletin - AG-407. North Carolina Agricultural Experiment Station, Raleigh. pp. 16-22.
8. De Vries, J., and T. L. Chow. 1978. *Hydrologic behavior of a forested mountain soil in coastal British Columbia.* Water Resources Research 14(5): 935-942.
9. Dindal, D. L. 1985. *Soil animals and soil fabric production: Facts and perceptions.* Quaestiones Entomologicae 12(4): 587-594.
10. Duncomb, D. R., W. E. Larson, C. E. Clapp, R. H. Dowdy, D. R. Linden, and W. K. Johnson. 1982. *Effects of liquid waste water sludge application on crop yield and water quality.* Journal of Water Pollution Control Federation 54: 1,185-1,193.
11. Edwards, C. A., and J. R. Lofty. 1977. *Biology of earthworms.* Chapman and Hall, London, England. 333 pp.
12. Edwards, W. M., L. D. Norton, and C. E. Redmond. 1988. *Characterizing macropores that affect infiltration into nontilled soil.* Soil Science Society of America Journal 52(2): 483-487.
13. Edwards, W. M., M. J. Shipitalo, L. B. Owens, and L. D. Norton. 1989. *Water and nitrate movement in earthworm burrows within long-term no-till corn fields.* Journal of Soil and Water Conservation 44(3): 240-243.
14. Edwards, W. M., M. J. Shipitalo, and L. D. Norton. 1988. *Contribution of macroporosity to infiltration into a continuous corn no-tilled watershed: Implications for contaminant movement.* Journal of Contaminant Hydrology 3: 193-205.
15. Fausey, N. R., and R. Lal. 1989. *Drainage and tillage effects on Crosby-Kokomo soil association in Ohio. 2. Soil temperature regime and infiltrability.* Soil Tech. 2: 373-385.
16. Gaiser, R. N. 1952. *Root channels and roots in forest soils.* Soil Science Society of America Proceedings 16(1): 62-65.
17. Germann, P. F., W. M. Edwards, and L. B. Owens. 1984. *Profiles of bromide and increased soil moisture after infiltration into soils with macropores.* Soil Science Society of America Journal 48(2): 237-244.
18. Greenland, D. J. 1981. *Soil management and soil degradation.* Journal of Soil Science 32(3): 301-322.
19. Hamblin, A. P., and D. B. Davies. 1977. *Influence of organic matter on the physical properties of some East Anglian soils of high silt content.* Journal of Soil Science 28(1): 11-22.
20. Hamilton, W. E., D. L. Dindal, C. M. Parkinson, and M. J. Mitchell. 1988. *Interaction of earthworm species in sewage sludge-amended soil microcosms: Lumbricus Terrestris and Eisenia Fetida.* Journal of Applied Ecology 25: 847-852.
21. Hartenstein, R. 1986. *Earthworm biotechnology and global biogeochemistry.* Advancement in Ecological Research 15: 379-409.
22. Hill, D. E., and J. Y. Parlange. 1972. *Wetting front instability in layered soils.* Soil Science Society of America Proceedings 36(5): 697-702.
23. Hillel, D. 1980. *Applications of soil physics.* Academic Press, New York, New York. 385 pp.
24. Hole, F. D. 1981. *Effects of animals on soil.* Geoderma 25: 75-112.
25. Kemper, W. D., R. C. Rosenau, and A. R. Dexter. 1987. *Cohesion development in disrupted soils as affected*

by clay and organic matter and temperature. Soil Science Society of America Journal 51(4): 860-867.

26. Kung, K.J.S. 1988. *Ground truth about water flow pattern in a sandy soil and its influences on solute sampling and transport modeling.* In P. J. Wierenga and D. Bachelet [editors] *International Conference on Validation of Flow and Transport Models, Unsaturated Zone.* Department of Agronomy and Horticulture, New Mexico State University, Las Cruces. pp. 224-230.
27. Lal, R., T. J. Logan, and N. R. Fausey. 1989. *Long-term tillage and wheel traffic effects on a poorly drained Mollic Ochraqualf in northwest Ohio. 1. Soil physical properties, root distribution, and grain yield of corn and soybeans.* Soil Tillage Research 14: 341-354.
28. Larson, W. E., and R. R. Allmaras. 1971. *Management factors and natural forces as related to compaction.* In *Compaction of Agricultural Soils.* American Society of Agricultural Engineers, St. Joseph, Michigan. pp. 367-427.
29. Lee, K. E. 1985. *Earthworms—their ecology and relationships with soils and land use.* Academic Press, New York, New York. 411 pp.
30. McKeague, J. A., C. A. Fox, J. A. Stone, and R. Protz. 1987. *Effects of cropping system on structure of Brookston clay loam in long-term experimental plots at Woodslee,* Ontario. Canadian Journal of Soil Science 67: 571-584.
31. Miyamoto, S., and J. Stroehlein. 1986. *Sulphuric acid effects on water infiltration and chemical properties of alkaline soils and water.* Transactions, American Society of Civil Engineers 29: 1,288-1,296.
32. Oades, J. M. 1984. *Soil organic matter and structural stability: Mechanisms and implications for management.* Plant and Soil 76: 319-337.
33. Petersen, A. E., E. E. Schulte, and F. H. Gustin. 1987. *Crop response and water pollution aspects of land application of bottom ash.* Agronomy Abstracts. American Society of Agronomy, Madison, Wisconsin. p. 32.
34. Piccolo, A., and J.S.C. Mbagwu. 1989. *Effects of humic substances and surfactants on the stability of soil aggregates.* Soil Science 147(1): 47-54.
35. Pikul, J. L., and R. R. Allmaras. 1986. *Physical and chemical properties of a Haploxeroll after 50 years of several residue management treatments.* Soil Science Society of America Journal 50(1): 214-219.
36. Rasmuson, A., and I. Neretnieks. 1986. *Radionuclide transport in fast channels in crystalline rock.* Water Resources Research 22(8): 1,247-1,256.
37. Rice, R. C., and R. S. Bowman. 1988. *Effect of sample size on parameter estimates in solute-transport experiments.* Soil Science 146(2): 108-112.
38. Richardson, J. L., W. T. Gunnerson, and J. F. Giles. 1986. *Influence of in site two-phase polymers on aggregate stabilization in various textured North Dakota soils.* Canadian Journal of Soil Science 67: 209-213.
39. Ritchie, J. T., D. E. Kissel, and E. Burnett. 1972. *Water movement in undisturbed swelling clay soil.* Soil Science Society of America Proceedings 36(6): 874-879.
40. Shaw, C., and S. Pawluk. 1986. *The development of soil structure,* Octolasion tyrtaeum, Aporrectodea turgida, *and* Lumbricus terrestris *in parent materials belonging to different textural classes.* Pedobiologia 29: 327-339.
41. Shipitalo, M. J., and R. Protz. 1988. *Factors influencing the dispersibility of clay in worm casts.* Soil Science Society of America Journal 52(3): 764-769.
42. Shipitalo, M. J., W. M. Edwards, W. A. Dick, and L. B. Owens. 1990. *Effect of initial storm size on macropore transport of surface-applied chemicals in blocks of no-till soil.* Soil Science Society of America Journal 54(6): 1,530-1,536.
43. Smettem, K.R.J., and N. Collis-George. 1985. *The influence of cylindrical macropores on steady-state infiltration in a soil under pasture.* Journal of Hydrology 79: 107-114.
44. Stone, J. A. 1987. *Compaction and the surface structure of a poorly drained soil.* Transactions, American Society of Agricultural Engineers 30(5): 1,370-1,373.
45. Theng, B.K.G. 1982. *Clay-Polymer interactions: Summary and perspectives.* Clays and Clay Minerals 30(1): 1-10.
46. Tisdale, J. M., and J. M. Oades. 1982. *Organic matter and water stable aggregates in soils.* Journal of Soil Science 33: 141-163.
47. Vorhees, W. B., W. W. Nelson, and G. W. Randall. 1986. *Extent and persistence of subsoil compaction caused by heavy axle loads.* Soil Science Society of American Journal 50(2): 428-433.
48. Zibilske, L. M. 1987. *Dynamics of carbon and nitrogen in soil during papermill sludge decomposition.* Soil Science 143(1): 26-33.

3

Soil Stresses Important to Management

P. H. Groenevelt

Interactions among structure, porosity, and water relations affect the physical condition of the soil at any time. Aggregate breakup is a major component in soil degradation and increased soil erosion. Therefore, the search for an effective stress variable that determines the onset of aggregate breakup is of prime importance.

Terzaghi (*4*) formulated the effective stress, σ', in saturated soils as the difference between the load (total or envelope) stress, P, and the liquid (pore water) pressure, p, as:

$$\sigma' = P - p1 \qquad [1]$$

where 1 is the unit tensor. The usefulness of σ' lies in its effectiveness to predict the onset of the massive (shear) failure of a soil body saturated with water, such as landslides. A useful effective-stress variable for unsaturated soils has yet to be presented.

In most cases involving unsaturated soils, a change in P causes a change in p that is now the (negative) hydrostatic pressure in the equilibrium solution, or the tensiometer pressure. This indicates that the porous medium is, in some way or another, deformable. There are various types of deformation, such as the massive (shear) failure of the unsaturated soil body, and aggregate rearrangement, deformation, and breakup. In the context of the structural degradation of agricultural soils, aggregate breakup is probably the most serious deformation.

If aggregate breakup (massive shear failure on the aggregate scale) is being considered, an effective-stress variable, which can predict the onset of this type of deformation should be defined.

Before attempting to define such a variable, some universal relations between the different pressures in porous media will be considered. These relations follow from fundamental thermodynamic principles and hold for all porous media in all states, regardless of texture, structural form, structural strength, or the type of deformation involved.

The first relation, derived by Groenevelt and Bolt (*2*), reads:

$$\left(\frac{\partial p}{\partial P}\right)_\theta = \left(\frac{\partial e}{\partial \theta}\right)_P \qquad [2]$$

where P represents the load pressure (as above), p represents the equilibrium liquid pressure (as above), e represents the void ratio (volume of voids per unit volume of solid material), and θ represents the moisture ratio (volume of liquid per unit volume of solid material).

The second relation, derived by Groenevelt

and Kay (1), reads:

$$\left(\frac{\partial p_s}{\partial P}\right)_\theta = e+1-\theta\left(\frac{\partial e}{\partial \theta}\right)_P \qquad [3]$$

were p_s is the equilibrium solid pressure. These fundamental relations come in differential form. Consequently, a definition of an appropriate effective stress will be sought in differential form. Subsequently, a suitable standard state will be chosen as a starting point for integration.

The Gibbs-Duhem equation:

$$\phi_s\, dp_s + \phi_w\, dp = dP \qquad [4]$$

where ϕ_s is the volume fraction of the solid material and ϕ_w is the volume fraction of the liquid, combined with the differential form of the Terzaghi equation 1:

$$d\sigma' = dP - dp \qquad [5]$$

yields, for saturated conditions:

$$d\sigma' = \phi_s d(p_s - p) \qquad [6]$$

This differential relation holds for $p \geq o$. It is now postulated that this relation also holds for $p<0$ as long as p is determined by the curvature of menisci and not by the thickness of waterfilms on the surfaces of the soil particles.

If a structured (aggregated) soil is considered, which is unsaturated, with all (or almost all) water present in the aggregates, and with a water content high enough so that, within the aggregates, p is determined by the curvature of the menisci, and equations 5 and 6 are applied within the aggregates, we may write:

$$d\sigma'_{agg} = dP_{agg} - dp \qquad [7]$$

and

$$d\sigma'_{agg} = \phi_{s,agg}\, d(p_s - p) \qquad [8]$$

Where $\phi_{s,agg} = \phi_s/\phi_{agg}$ (i.e., the volume fraction of the solid material within the aggregates), ϕ_{agg} is the volume fraction of the aggregates (volume of aggregates per unit volume bulk), $P_{agg} \equiv P/\phi_{agg}$ (the load pressure per unit area occupied by aggregates) and $\sigma'_{agg} \equiv \sigma'/\phi_{agg}$ (the effective stress within the aggregates). This variable, σ'_{agg}, appears to be the effective stress variable, which is able to predict massive shear failure within aggregates (i.e., the onset of aggregate breakup).

When the solid material is surrounded by liquid, of which the equilibrium pressure is dictated by mechanical effects, and not by osmotic effects, then the differential of the intra-aggregate effective stress can be equated to the product of the volume fraction of the aggregates and the differential of the difference between p_s and p.

For aggregate beds in which the volume fraction of the aggregates is constant, equation 8 is identical to equation 6. Equation 8 may now be differentiated with respect to P.

$$\frac{d\sigma'_{agg}}{dP} = \phi_{s,agg}\left(\frac{dp_s}{dP} - \frac{dp}{dP}\right) \qquad [9]$$

Using equations 2 and 3, and

$$(e + 1) = 1/\phi_s \qquad [10]$$

equation 9 may be rewritten as:

$$\phi_{agg}\text{-}\frac{d\sigma'_{agg}}{dP} = 1 - \left(\frac{\theta+1}{e+1}\right)\left(\frac{\partial \varepsilon}{\partial \theta}\right)_P \qquad [11]$$

Using equation 2 again and integrating equation 11 with respect to P from the standard state, where P=0, p=0, and $\sigma'_{agg}=0$, to the state under consideration, one obtains:

$$\int_{\sigma'_{agg}=0}^{\sigma'_{agg}} \phi_{agg} d\sigma'_{agg} = P - \int_{p=0}^{p}\left(\frac{\theta + 1}{e + 1}\right)dp \qquad [12]$$

For saturated conditions, the volume fraction of the aggregates, ϕagg, is constant, and the variable

$$(\theta + 1)/(e + 1) = \phi_s + \phi_w \qquad [13]$$

has the constant value 1. Equation 12 then reduces to $\phi_{agg}\ \sigma'_{agg} = \sigma' = P - p$, which is identical to Terzaqhi's equation.

Under very dry conditions, p is determined by osmotic, electrical double-layer and London-Van der Waals forces, and the influence on the intergranular pressure disappears. By deleting the contribution made by p from equations 8 and 9, the integrand in the last term of equation 12 becomes

$$\theta/(e + 1) = \phi_w \qquad [14]$$

Thus, under completely dry conditions and constant ϕ_{agg},

$$\sigma'_{agg} = P/\phi_{agg} \quad [15]$$

For the intermediate, unsaturated state, in which all the water is present within the aggregates and all the air is present in the interaggregate pore spaces so that

$$(\theta + 1)/(e + 1) = \phi_{agg} \quad [16]$$

from equation 12, for constant ϕ_{agg},

$$\sigma'_{agg} = (P/\phi_{agg}) - p \quad [17]$$

During the process of increasing the load, P, on a bed of aggregates, the value of p usually increases, but it occasionally decreases. When p decreases upon loading, both the direct contribution of P and the indirect influence of P via p, increase the value of the intergranular pressure, σ'_{agg}. Therefore, the aggregates are strengthened to resist shear stresses and breakup. However, when p increases upon loading, the indirect influence of P via p lowers the intergranular pressure, thus weakening the aggregates in their resistance to shear stresses.

If, during the process of loading a bed of aggregates, p first decreases and subsequently increases during the same loading process it is likely that, at the point of reversal of the direction in which p changes, aggregate breakup will occur, as was suggested by Larson and Gupta (*3*).

REFERENCES CITED

1. Groenevelt, P. H., and B. D. Kay. 1981. *On pressure distribution and effective stress in unsaturated soil.* Canadian Journal of Soil Science 61: 431-443.
2. Groenevelt, P. H., and G. H. Bolt. 1972. *Water retention in soil.* Soil Science 113(4): 238-245.
3. Larson, W. E., and S. C. Gupta. 1980. *Estimating critical stress in unsaturated soils from changes in pore water pressure during confined compression.* Soil Science Society of America Journal 44(6): 1,127-1,132.
4. Terzaghi, K. 1925. *Erdbaumechanik auf bodenphysikalischer Grundlage.* Deuticke, Vienna.

4

Soil Compaction Mechanisms and Their Control

R. M. Cruse and S. C. Gupta

Compaction is an unintended consequence of many farming practices. To some extent, technology has created a compaction dilemma. Early spring planting generally has greater crop production potential than later planting. Early spring farming operations frequently take place under wet soil conditions, which favor soil structural breakdown and compaction from farming activities. With present-day high-power, multiple-axle-driven tractors, multiple tire mounts on each axle, and/or "flotation" tire designs, farmers can conduct field operations under wetter soil conditions than they could in the past. The conflict between soil structural degradation and compaction-induced yield reductions as well as early planting yield enhancement continues to increase as the technological capability for farming wet soils expands. Because of this technology, there exists an increasing need to understand how structured agricultural soils respond to load applications across a broad soil-water content range.

According to McKibben (*20*), between 1948 and 1968 the average tractor weight increased from 2.7 tons to 4.5 tons. At present, the average is 6.8 tons, with larger units weighing more than 22.4 tons. Gill (*13*) estimated an annual loss of $1.18 billion in the United States from decreased yields and increased energy costs caused by soil compaction during field tillage.

Compaction Processes

During compaction, large interaggregate pores are lost first (i.e., aggregates are pressed closer together removing interaggregate pore space). The next compaction stage involves aggregate breakdown, a process that leads to massive structure (i.e., integrity of individual aggregates is lost). An example of the changes in pore structure due to an externally applied stress is shown in figure 1 and table 1 (*15*). With an increase in applied stress (bulk density), there is (a) a reduction in the total volume of soil, as indicated by water retention at the highest potential (pounds per square inch = -0.1 kilopascals); (b) a reduction in the proportion of large pores (water retention at soil matric potential ranges from -0.1 to -1.0 kilopascals); and (c) an increase in the proportion of smaller pores (water retention at the soil matric potential of ≤ -1.0 kilopascals). The trend toward flattening of the sigmoidial shape of the water retention curve (Figure 1) with increased compaction indicates a trend toward dominance of smaller size pores (corresponding to the lowest matric potential). Of great importance to management is that once aggregate breakdown occurs and small pores dominate, regeneration of favorable structural conditions may take several years (*16*).

Compaction Effects on Soil Physical/Hydrological Properties

Soil is composed of aggregates surrounded by large pores. Aggregates are high-bulk-density soil units containing small pores that enhance water retention and nutrient storage. The relatively large pore spaces between aggregates promote rapid water infiltration, water drainage, air exchange, and root growth. As shown above, during compaction, aggregates are pressed closer together, resulting in a loss of large pores. If compaction forces are sufficiently large, aggregate breakdown occurs. Therefore, small pores dominate in compacted soil, resulting in slow water infiltration, slow water drainage, reduced aeration, and increased resistance to plant root growth.

Akram and Kemper (*1*) showed that, for a given water content at the time of soil compaction, infiltration rates in laboratory-prepared soil cores decrease with an increase in compaction stress. Earlier, Davies and associates (*6*) showed that the field infiltration in a clay loam soil decreased from 71.0 to 7.6 and 2.7 centimeters per hour when untreated land was subjected to applied mechanical loads of 1,710 and 3,140 kilograms, respectively. Similarly, Bowen (*3*) observed that air permeability of a loamy fine sand at water contents between 8 to 10 grams per gram decreased from a maximum 4.5 x 10^{-7} to 1.3 x 10^{-7} square centimeters when subjected to an applied stress of 34.5 kilopascals. When considering the effects of compaction on gas diffusion, Currie (*5*) concluded that, with increased compaction, the contribution of large pores (between aggregates) to gas diffusion

Table 1. Pore size distribution in a Nicollet clay loam at four levels of soil compaction.

Applied Stress (KPa)	Pore Volume (%) Pore Size (μm)			
	>150	15-150	1-15	<1
5.8	21.1	35.7	10.9	32.3
14.8	12.5	33.9	13.6	40.0
45.0	9.4	23.7	19.2	47.7
113.8	6.2	12.2	16.7	64.9

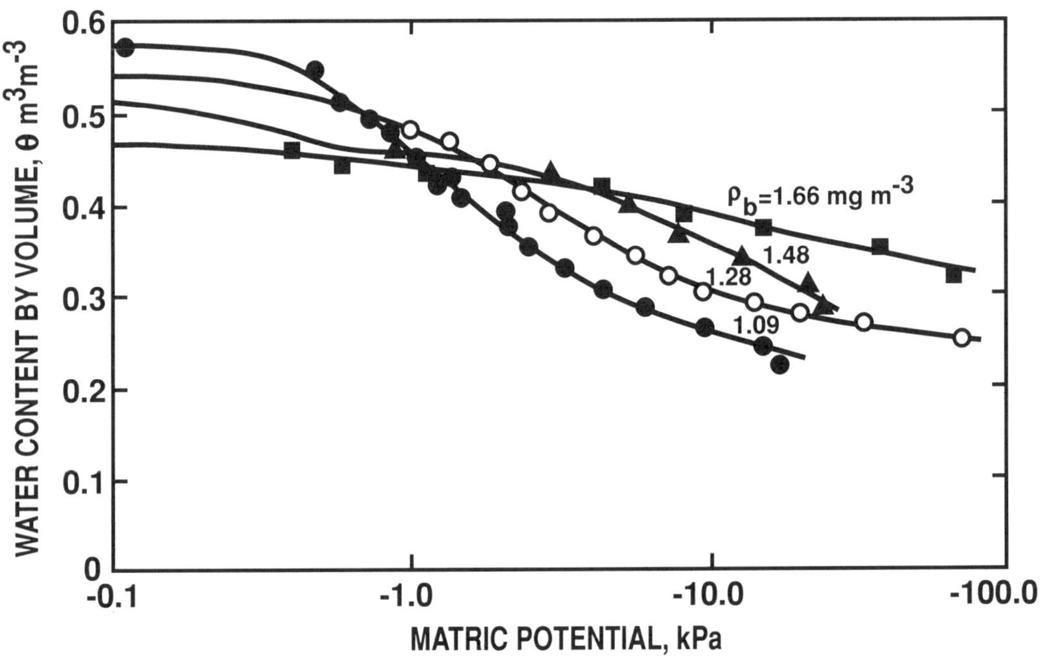

Figure 1. Water retention characteristics of Nicollet clay loam at four bulk densities.

becomes increasingly small and a greater proportion of diffusion occurs through the pores in the aggregates. Compaction affects soil mechanical properties of subsequently tilled soils by increasing the bulk density, clod population, clod density, and shear strength (*11*). Voorhees and associates (*23*) found that the proportion of large aggregates increased when silty clay loam soil was subsequently tilled after five years of soil compaction. Also, clod density and crushing strength was higher for clods from tilled wheel tracked areas than from tilled nontracked areas.

Predicting when and under what conditions compaction will occur is difficult at best. Soil type, farming practices, and farming equipment change with time and location; thus, compaction potential varies with both time and location. However, one consistency exists: if soil aggregate breakdown occurs during the load application, soil physical-property problems are likely to follow. Therefore, the fundamentals of identifying soil conditions and applied loads that minimize excessive compaction are based on understanding the stability limits of soil aggregates/peds at various applied loads and water contents.

Aggregate Stability Tests

Dexter and Kroesbergen (*8*) indicated that aggregate tensile strength is the most useful measure of soil-aggregate physical condition. In this test, aggregates are pressed between parallel plates until aggregate failure occurs, and aggregate tensile strength or energy required to crush aggregates is measured. This technique allows aggregate testing at selected initial water contents or water potentials, although most consistent results occur with dry aggregates (*8*). This technique is limited in that the aggregate must be nearly spherical at failure. At higher soil water contents it is difficult to define the stress when the aggregate breaks because of plastic flow.

Wet aggregate stability has also frequently been used to assess soil management impacts on soil aggregation. Soil aggregates are submerged and gently sieved under water to characterize the aggregate resistance to breakdown (*18*). Although the procedure has been standardized, the water conditions in the test are considerably different from those under most field conditions during compaction. The wet-sieve test forces the soil water matric potential of the aggregate to zero, whereas matric potentials during compaction are generally much lower. Surface sealing or puddling processes during intense rainstorms or under flood irrigation are more closely related to the physical processes that occur in the wet aggregate stability test. This test is easy to use and very sensitive for identifying treatment effects on soil aggregate development. However, it is not suitable for characterizing soil resistance to compaction because of differences in soil water potential between the test and that occurring at the time of compaction.

A third method that characterizes aggregate and/or soil resistance to compaction is the compression test. Through a series of uniaxial load applications, the relationship between load application and resulting soil bulk density is obtained. Examples of these plots are provided in figure 2. This test more closely duplicates, than do other tests, the soil conditions under which compaction occurs and is well suited for soils with various water contents. This test is limited in that it is an indirect measure of aggregate stability. Additionally, load applications during the compression test normally occur for time periods sufficiently long that soil subsistence ceases or nearly ceases. This may take from several minutes to hours.

Compaction Effects on Pore Water Potential

When conducting uniaxial compression tests, Larson and Gupta (*19*) observed a fundamental and very important soil response to load applications. As force was applied to ground and sieved soil, water potential changed during the test. It first decreased (became more negative), and then increased to zero. The extent of water potential change was dependent on soil water content at test initiation and soil type. Figure 2 exemplifies the results of this study. Larson and Gupta (*19*) suggested that the stress that corresponds to minimum pore water potential serves as an indicator of aggregate packing during the compaction process. This observation has been supported by thermodynamic analysis of aggregate packing during compaction (*14*) and visual observation of aggregate breakdown in thin sections (*7*). Gupta and associates (*15*) provided a detailed review of water potential, aggregate breakdown, and compaction interactions.

To recognize the importance of change in soil water potential during compression, the role of soil water potential on soil resistance

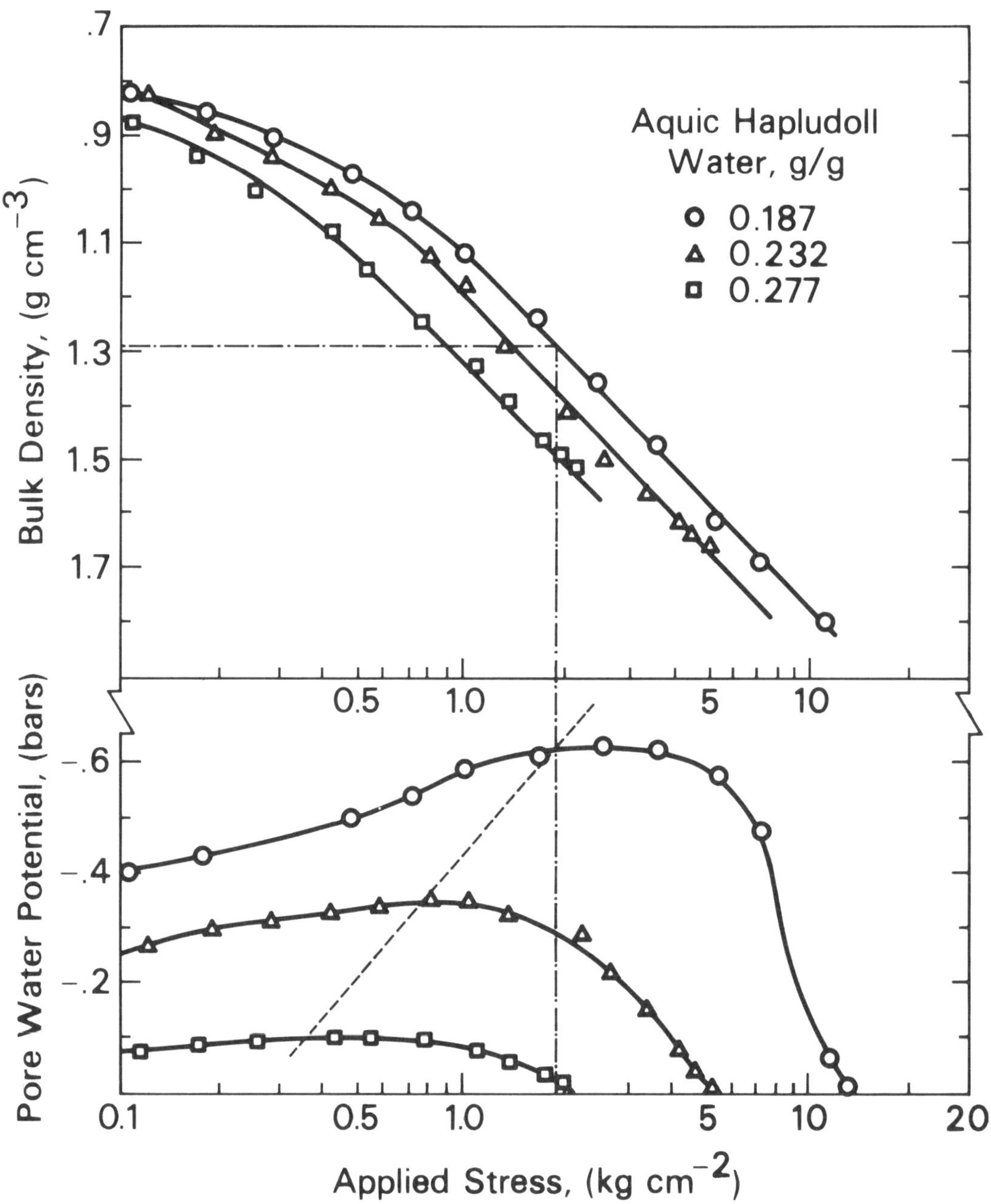

Figure 2. Soil bulk density and pore water potential as influenced by applied stress (load).

to compaction needs to be considered. Soil shear strength, as an indicator of this resistancc, is inversely related to soil water matric potential. The relationship between matric potential and soil shear strength in nonlinear, except at saturation (*24*). Additionally, this relationship seems to be unique for each soil (*24*). This relationship is complex, yet very important, particularly if aggregate stability (or resistance to compaction) is affected by matric potential changes, as in bulk soil.

Limited direct (*2*) and indirect (*12*) evidence suggests that aggregate strength is affected by matric potential changes similarly to that of bulk soil. Braunack and associates (*4*), Dexter and Tanner (*9*), and McMurdie and Day (*21*) indicated that soil water content and/or potential have a major effect on aggregate tensile strength. A functional relationship between soil water matric potential and soil aggregate strength has not been evaluated. To appropriately evaluate structured soil's susceptibility to compaction, aggregate stability must be evaluated at water potentials comparable to those occurring in the field. Aggregate stability tests using very high, very low, or unknown water potentials do not reasonably simulate the physical processes occurring in the field during compaction.

The Larson and Gupta (*19*) approach is preferable to other tests and appropriately addresses aggregate breakdown during soil compression. However, it may not suitably address aggregate breakdown during compaction. Compaction forces are normally applied and removed very rapidly, compared with the static load applications applied in this test. The duration of load application, at least as indicated by tractor wheel speed, does affect soil compaction, as indicated by results of Horn and associates (*17*), Dexter and Tanner (*10*), and Vomocil and associates (*22*), among others. Equating results of static load applications, as is done with the compression tests, to results occurring with rapid loading and unloading is risky at best.

There are two major contributions to compaction research that are needed at this time. First, matric potential changes occurring with rapid soil loading and unloading, both during field and laboratory compression, need to be addressed. Second, the role of matric potential change on aggregate stability needs to be more clearly defined.

Summary

Severe compaction problems in structured soils are associated with compaction forces that cause aggregate breakdown. The most prominent challenge facing researchers is to determine the force that aggregates can sustain without failing. This challenge would be relatively easy to meet if aggregate strength was static (i.e., remained constant) during the compaction process. This does not seem to be the case.

Soil water potential affects aggregate strength and, because soil water potential seems to change during compaction, aggregate strength will also vary during the compaction process. In addition to understanding aggregate stability as a function of matric potential, an understanding is needed of the water potential change that occurs during the compaction process and the effect of this variable on soil aggregate stability.

REFERENCES CITED

1. Akram, M., and W. D. Kemper. 1979. *Infiltration of soils as affected by the pressure and water content at the time of soil compaction.* Soil Science Society of American Journal 43: 1,080-1,086.
2. Benjamin, J. G., and R. M. Cruse. 1985. *Measurements of shear strength and bulk density of soil aggregates.* Soil Science Society of American Journal 49: 1,248-1,251.
3. Bowen, H. D. 1966. *Measurement of edaphic factors for determining planter specifications.* Transactions, American Society of Agricultural Engineers 9: 725-735.
4. Braunack, M. V., J. S. Hewitt, and A. R. Dexter. 1979. *Brittle fracture of soil aggregates and the compaction of aggregate beds.* Journal of Soil Science 30: 653-667.
5. Currie, J. A. 1984. *Gas diffusion through soil crumbs: the effects of compaction and wetting.* Journal of Soil Science 35: 1-10.
6. Davies, D. B., J. B. Finney, and S. J. Richardson. 1973. *Relative effects of tractor weight and wheelslip in causing soil compaction.* Journal of Soil Science 24: 399-409.
7. Day, P. R., and G. G. Holmgren. 1952. *Microscopic changes in soil structure during compression.* Soil Science Society of America Proceedings 16: 73-77.
8. Dexter, A. R., and B. Kroesbergen. 1985. *Methodology for determination of tensile strength of soil aggregates.* Journall of Agricultural Engineers Research 31: 139-147.
9. Dexter, A. R., and D. W. Tanner. 1973. *The response of unsaturated soils to isotropic stress.* Journal of Soil Science 24: 491-502.

10. Dexter, A. R., and D. W. Tanner. 1974. *Time dependence of compressibility for remoulded and undisturbed soils.* Journal of Soil Science 25: 153-164.
11. Flocker, W. J., J. A. Vomocil, and M. T. Vittum. 1958. *Response of winter cover crops to soil compaction.* Soil Science Society of America Proceedings 22: 181-184.
12. Francis, P. B., and R. M. Cruse. 1983. *Soil water matric potential effects on aggregate stability.* Soil Science Society of America Journal 47: 478-581.
13. Gill, W. R. 1971. *Economic assessment of soil compaction.* In K. K. Barnes, W. M. Carleton, H. M. Taylor, R. I. Throckmorton, and G. E. Vanden Bergs [editors] *Compaction of Agricultural Soils.* Monograph No. 1. American Society of Agricultural Engineers, St. Joseph, Michigan. pp. 431-458.
14. Groenevelt, P. H., and B. D. Kay. 1981. *On pressure distribution and effective stress in unsaturated soils.* Canadian Journal of Soil Science 61: 431-443.
15. Gupta, S. C., P. P. Sharma, and S. A. DaFranchi. 1989. *Compaction effects on soil structure.* Advances in Agronomy 42: 311-338.
16. Hakansson, I., W. B. Voorhees, and H. Riley. 1988. *Vehicle and wheel factors influencing soil compaction and crop response in different traffic regimes.* Soil Tillage Research 11: 239-282.
17. Horn, R., P. S. Blackwell, and R. White. 1989. *The effect of speed of wheeling on soil stresses, rut depth and soil physical properties in an ameliorated red-brown earth.* Soil Tillage Research 13: 353-364.
18. Kemper, W. D., and R. C. Rosenau. 1986. *Aggregate stability and size distribution.* In Arnold Klute and associates [editors] *Methods of Soil Analysis.* Part I. Agronomy 9: 425-442. American Society of Agronomy, Madison, Wisconsin.
19. Larson, W. E., and S. C. Gupta. 1980. *Estimating critical stress in unsaturated soils from changes in pore water pressure during confined compression.* Soil Science Society of America Journal 44: 1,127-1,132.
20. McKibben, E. G. 1971. *Introduction.* In K. K. Barnes, W. M. Carleton, H. M. Taylor, R. I. Throckmorton, and G. E. Vander Bergs [editors] *Compaction of Agricultural Soils.* Monograph No. 1. American Society of Agricultural Engineers. St. Joseph, Michigan. pp. 3-6.
21. McMurdie, J. L., and P. R. Day. 1958. *Compression of soil by isotropic stress.* Soil Science Society of America Proceedings 22: 18-21.
22. Vomocil, J. A., E. R. Fountaine, and R. E. Regenato. 1958. *The influence of speed and drawbar load on the compacting effect of wheeled tractors.* Soil Science Society of America Proceedings 22: 178-180.
23. Voorhees, W. B., R. A. Young, and L. Lyles. 1979. *Wheel traffic consideration in erosion research.* Transactions, American Society of Agricultural Engineers 22: 786-790.
24. Yong, R. N., and B. P. Warkentin. 1966. *Granular soil strength and strength theories.* In Gene Norbby [editor] *Introduction to Soil Behavior.* Macmillan Company, New York, New York. pp. 235-280.

5

Soil Erosion: Processes, Impacts, and Prediction

W. J. Elliot, G. R. Foster, and A. V. Elliot

Soil erosion occurs naturally. The detachment and transportation of soil particles has helped to shape our landscape and develop soil profiles used in such important human activities as crop production and the preparation of building sites. These activities usually accelerate erosion, however, causing excessive loss of the soil's productive capacity and, in some situations, off-site environmental damage.

Practical measurement of soil erosion is difficult. Erosion is a diffuse process that occurs at relatively low rates and widely varying rates from year to year and from location to location. Technology for estimating rates of soil erosion has emerged as a major tool to overcome these difficulties. Prediction technology currently is a collection of mathematical expressions used to estimate rates of soil erosion as functions of independent variables with values that represent a given set of conditions.

Soil Erosion Prediction as a Tool

Erosion prediction has been most widely used as a tool to guide conservation planning. In this application an estimate of erosion at a particular site is made based on the site characteristics. If the computed erosion rates exceed a rate accepted as tolerable for the given soil, alternative erosion practices are proposed, and a practice is chosen that gives a computed erosion rate equal to or less than the tolerable erosion rate. Used in this manner, erosion prediction is a powerful tool for selecting conservation practices tailored to particular site conditions.

In the last decade, erosion prediction has become widely used to assess erosion on a national scale. For example, soil erosion was estimated at more than one million sample points on nonfederal land in the United States for the 1982 National Resources Inventory. These erosion estimates were compiled in many ways to show the distribution of erosion by political boundary, geographic region, soil type, land use, and land classification. The results of these analyses guided development of public policy embodied in the Food Security Act (FSA) of 1985.

One FSA title requires that farmers apply conservation measures to remain eligible to participate in certain government programs. Erosion prediction technology is used to help identify areas needing conservation measures and to determine what measures should be applied on specific fields and farms to meet the requirements of FSA.

The adequacy and effectiveness of soil erosion control measures, including those influenced by FSA, depends to a significant degree on the adequacy of the soil erosion prediction technologies used.

Development of Soil Erosion Prediction Technology

Erosion prediction technology can be characterized as empirically based, process based, or a combination of the two. Either type of erosion prediction technology requires large amounts of data. The universal soil loss equation (USLE) (*48*) and the wind erosion equation (WEQ) (*49*) represent empirically based technology. The USLE, which predicts sheet and rill erosion by water, and the WEQ, which predicts erosion by wind, are by far the most widely used methods for predicting soil erosion. Both equations were used in the 1982 National Resources Inventory, and both equations have been used for more than two decades in conservation planning. The USLE and the WEQ are the prediction technologies being used by agencies implementing FSA. Research and development are underway to replace these empirical equations with a process-based methodology (*13*).

Empirically Based Erosion Prediction Technology. The conceptual basis for the USLE and WEQ is that erosion is a function of the erosivity of the climate, inherent soil erodibility, topography, and man's activities. Erosion is computed using indices that represent each of the major factors.

Erosion prediction was first used as a practical tool in the 1940s; Zingg's equation related sheet and rill erosion to slope length and steepness (*51*). Smith and Whitt (*36*) soon added cropping and management factors to the equation. In 1947 Musgrave (*30*) described an erosion equation developed by a committee that he chaired; that committee had reviewed the data base on soil erosion that existed at that time.

The universal soil loss equation. The conceptual framework for the USLE was presented by Cook (*7*) in 1936. In the early 1950s Wischmeier and his associates began to assemble and analyze the data available; they developed the USLE in the late 1950s and early 1960s (46). That equation remains unchanged today:

$$A = RKLSCP \qquad [1]$$

where A is average annual soil loss, R is rainfall erosivity, K is the soil erodibility factor, L is the slope length factor, S is the slope steepness factor, C is the cover management factor, and P is the supporting practice factor (*48*).

Over the next few years, Wischmeier and others continued their development of the USLE, which resulted in a major update of the equation in 1978 (*48*). The rainfall erosivity factor, a function of climate, originally was defined for all areas east of the Rocky Mountains. Estimates have since been made for mountainous areas also (*48*).

The soil erodibility factor can be estimated from soil properties by the following regression equation:

$$K^{1} = 2.8 \times 10^{-7} M^{1.14} (12 - a) + 4.3 \times 10^{-3} (b - 2) + 3.3 \times 10^{-3} (c - 3) \qquad [2]$$

where M is the particle size parameter [(% silt + % very fine sand) *x* (100 - % clay)], a is the percentage of organic matter, b is the soil structure code (very fine granular, 1; fine granular, 2; medium or coarse granular, 3; blocky, platy, or massive, 4), and c is the profile permeability class (rapid, 1; moderate to rapid, 2; moderate, 3; slow to moderate, 4; slow, 5; very slow, 6).

The slope length factor (*48*) is found from the slope by the relationship:

$$L = \left(\frac{1}{22}\right)^{m} \qquad [3a]$$

where m is 0.2 for $s < 1$ percent, 0.3 for 1 percent $< s < 3$ percent, 0.4 for 3.5 percent $< s < 4.5$ percent, and 0.5 for $s > 5$ percent; 1 is the slope length, and s is the field slope (percent).

McCool and associates (*25*) recommended that m be defined as follows:

$$m = \frac{\text{rill erosion from slope}}{\text{rill erosion} + \text{interrill erosion}} \qquad [3b]$$

where rill and interrill erosion are calculated by equations similar in form to 7 and 8, combined with topographic observations.

A revised slope steepness factor was developed by McCool and associates (*26*). This factor varies with slope length. The set of equations used to define the slope steepness factor are as follows:

For slopes less than 4 meters long:

$$S=3.0[(\sin(\theta)]^{0.8}+0.56 \quad [4a]$$

For slopes greater than 4 meters long, and s < 9 percent:

$$S=10.8 \sin(\theta)+0.03 \quad [4b]$$

For slopes greater than 4 meters long, and s ≥ 9 percent:

$$S=16.8 \sin(\theta)-0.50 \quad [4c]$$

where θ is the field slope in degrees $[=\mathrm{Tan}^{-1}(s/100)]$

The cover management factor, C, depends on crop growth rate in relation to the erosivity variation in the climate. Wischmeier and Smith (*48*) described how to calculate the C factor, and many Soil Conservation Service offices have developed C factors to suit local crops and climates (*4, 37*). Currently, the Agricultural Research Service is updating the USLE using a computer program (*32*) that will allow for more detailed consideration of the C factor (*22*).

The supporting practice factor, P, reflects the use of contours, contour stripcropping, and terraces on soil erosion. This factor can vary from as low as 0.05 for terraces with underground outlets to 1.0 for no practice (17).

Because the USLE is an empirical equation, its development required a large, extensive, and comprehensive data base. Fortunately, soil conservationists had the vision to establish a network of erosion experiment stations in the 1930s that produced a wealth of data for a wide range of climatic, soil, topographic, and land use conditions across the United States east of the Rocky Mountains. In the early 1930s, 10 federal erosion experiment stations were established at locations including Bethany, Missouri; Clarinda, Iowa; Statesville, North Carolina; Tyler, Texas; and Zanesville, Ohio. Each of these stations had a common experimental design that produced a unified data base that allowed the isolation, definition, and determination of the major factors affecting erosion. These stations operated for about 10 years, with some of the stations, such as Clarinda, Iowa, operating for a much longer period.

In the early 1940s several state agricultural experiment stations established experiments patterned after those at the federal stations. Some of the states having experiments included Arkansas, New York, New Jersey, Maine, and Tennessee.

As the USLE was being developed in the late 1950s and early 1960s, a need to have current data was recognized, and several new federal experimental sites were established at Tifton, Georgia; Presque Isle, Maine; and Morris, Minnesota; among others (48).

Typical experiments at these stations used field plots about 2 meters wide and 22 meters long to evaluate soil erodibility as well as cropping and management practices. These stations also included small watersheds, about 2 hectares in size, for evaluating such practices as contouring, stripcropping, and terracing. Data from these studies were collected under natural rainfall conditions. Usually, 5 to 10 years of data were required to arrive at conclusions with reasonable confidence.

Conducting erosion research using natural rainfall or wind is costly and time-consuming. In the 1950s, portable rainfall simulators and wind tunnels were developed to accelerate the collection of experimental data under controlled conditions. Data from wind tunnels were important for deriving the WEQ (*6*). In the case of the USLE, data produced by rainfall simulators were used to derive the soil erodibility nomogram and cover management factors for conservation tillage and rangeland conditions (*48*). Rainfall simulators and portable wind tunnels have been transported and used in experiments all over the United States and in many other countries. Similarly, use of rainfall simulators has been an important tool in the laboratory.

Although the USLE is the standard tool for estimating erosion, other technology is available for predicting erosion by water. These methods are for individual storms rather than for the average annual loss. The runoff terms are computed with a hydrologic model that uses the SCS curve number method or a more fundamental approach, such as the Green-Ampt infiltration relationship, to compute the portion of rainfall that becomes runoff during an individual storm. Much of this technology uses a modification of the USLE by replacing the erosivity factor with one that includes runoff terms rather than, or in addition to, the rainfall term. One widely used model is (*42*):

$$Y = 11.8\ (Qq)^{0.56}\ KLSCP \qquad [5]$$

where Y is the single storm sediment yield; Q is the storm runoff; q is the peak storm discharge; and K, L, S, C, and P are the same as in equation 1.

Frequently sediment "yield" from a hillslope or watershed, rather than annual sediment "loss" from a hillslope, is predicted. Sediment yield is usually lower than sediment loss and may be related to sediment loss by a delivery ratio which usually varies with size of watershed (*34*).

The Wind Erosion Equation. The WEQ, developed by Woodruff and Siddoway (*49*), was based much on the conceptual thinking of Chepil (*5*). It too has undergone continued updating. The current relationship is in the functional form:

$$E = f(I, K, C, L, V) \qquad [6]$$

where E is the estimated average annual soil loss (mass/area), I is the soil erodibility index, K is the ridge roughness factor, C is the climate factor, L is the unsheltered length of eroding field factor, and V is the vegetative cover factor. The relationships among the parameters are not a simple product, but rather are defined by a set of tables, graphs, and nomograms (*38*).

Drawbacks to Empirical Erosion Prediction Technology. Conducting erosion research using natural rainfall or wind is costly and time-consuming. In the 1950s portable rainfall simulators and wind tunnels were developed to accelerate the collection of experimental data under controlled conditions. Data from wind tunnels were important to development of the WEQ. In the case of the USLE, data produced by rainfall simulators were used to derive the soil erodibility nomogram and cover manage-

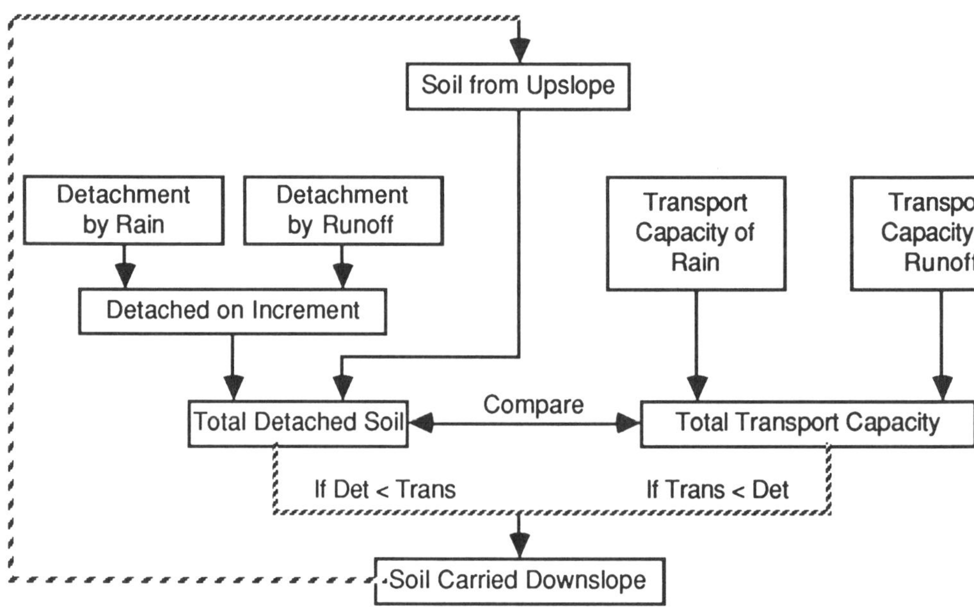

Figure 1. The conceptual water erosion model proposed by Meyer and Wischmeier (29).

ment factors for conservation tillage and rangeland conditions.

Even though the addition of a runoff erosivity term significantly improved the USLE, the equation still does not fully represent the data and the relationship among variables that knowledge of erosion processes indicates should be represented in erosion prediction technology (*27, 45*). Although runoff variables can be introduced into the USLE in its erosivity factor, other variables in the USLE also are functions of runoff. Introducing runoff into all of the USLE variables is complex, and the form of the equation does not allow it to represent erosion processes adequately. A similar argument can be made for the inadequacy of the WEQ. If significant improvements are to be made in erosion prediction technology, the technology must be based on fundamental erosion processes rather than on empirically based equations, like the USLE and WEQ.

Process-Based Erosion Prediction Technology. The fundamental erosion processes are detachment, transport, and deposition by the erosive agents—raindrop impact and surface runoff for erosion by water, and wind for erosion by wind. In the case of wind erosion, the processes are more frequently considered to be creep, saltation, abrasion, and suspension (38). The main concepts used in process-based prediction for erosion by water were presented by Ellison (*11*). Meyer and Wischmeier (*29*) implemented Ellison's concepts in a mathematical simulation model (Figure 1) that showed how these relationships could be used to compute erosion and deposition along nonuniform slopes. This development was a significant advancement because the USLE applied only to uniform slopes. Later, Foster and Wischmeier (*18*) showed how the USLE could be applied to the eroding portion but not the depositional portion of nonuniform slopes.

Also, Negev (*31*) showed in the 1960s how relationships for fundamental erosion processes could be combined with relationships for fundamental hydrologic processes in a hydrologic simulation model to compute erosion. Both Meyer and Wishmeier's method and Negev's method were computationally intensive. Because computers were not widely available at the time, process-based modeling remained a research tool only for several years. Within the last five years computers have become sufficiently powerful and widely available to make feasible erosion prediction based on fundamental hydrologic and erosion processes.

In the 1970s, nonpoint-source pollution became a major concern (*33*). Because sediment itself is a pollutant and a carrier of pollutants, much attention was focused on soil erosion. However, instead of being concerned about erosion rates on the slopes the concern was for sediment delivered from fields and the characteristics of that sediment. Direct application of the USLE was technically insufficient. Many analyses of nonpoint-source pollution were conducted by a relatively few people with access to computers instead of by field personnel doing most of the conservation planning for controlling erosion to maintain productivity.

This situation led to the development of several detailed erosion and sediment delivery prediction models, including CREAMS (*21*) AGNPS (*50*), ANSWERS (*3*), and SEDIMOT II (*44*). Some of these models, following Meyer and Wischmeier's lead, were able to describe spatial variations in erosion and deposition over the landscape. In many cases the models used modifications of the USLE to compute detachment because parameter values were not available for process-based detachment relationships.

With an obvious need to consider spatial variability, much attention was focused on representing complex topography. Some of the schemes included a square grid system (AGNPS and ANSWERS), a contour-streamube representation, and a simple stream order-representative profile approach (SEDIMOT II).

Lumped parameters, combining rill and interrill erosion or runoff and erosion, versus distributed parameters was a major issue during this period. One problem with lumped parameters was the distortion in parameter values, which were determined by calibration. Values determined by calibration often were so highly varied and nonsensical that they could not be related to measurable properties in the watershed represented by the model.

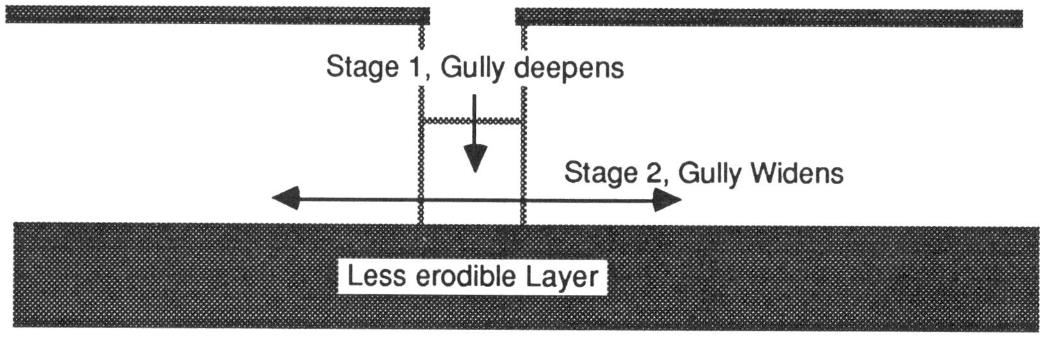

Figure 2. Erosion process in an ephemeral gully.

The rill-interrill concept dealt with this problem by isolating erosion processes so they could be measured.

The rill-interrill concept puts erosion processes into a spatial context. Runoff tends to concentrate in many small flows across natural surfaces. If these flows cause erosion and the erosion is allowed to continue, the flow will erode incised rills into the soil surface. These flow concentrations, regardless of whether they develop into rills, are called rill areas, and the areas between the rills are called interrill areas. The processes that predominate in the rill areas are detachment, transport, and deposition by flow. Raindrop impact processes predominate on the interrill areas. Impacting raindrops detach soil particles, and the detached particles are transported to the rills by the thin surface flow. The transport capacity of this flow is greatly enhanced by raindrop impact. In some cases, some transport by raindrop impact or splash, may occur (*15*).

Rill erosion rates are believed to be related to the hydraulic shear of the water flowing in the rill. Rill erosion can be estimated by the relationship (*23*):

$$D_r = K_r(\tau - \tau_c)\left(1 - \frac{G_s}{T_c}\right) \qquad [7]$$

where D_r is the rill erosion rate, K_r is the rill erodibility of the soil, τ is the shear force of the water flowing in the rill, τ_c is the soil's critical shear below which detachment is negligible, G_s is the amount of sediment in transport, and T_c is the transport capacity of the rill. K_r and τ_c are soil properties. Computer erosion models of the form presented in figure 1 are necessary to evaluate equation 7 because D_r is the rate of change in G_s, and τ and T_c vary with the amount of flow in the rill, which is a function of position on the slope.

Interrill erosion is believed to be a function of soil properties, rainfall intensity, and slope. One common model for interrill erosion (*9, 14*) is

$$D_i = K_i\ I^2\ S_f\ C_i \qquad [8]$$

where D_i is the interrill erosion rate, K_i is the interrill erodibility of the soil, I is the rainfall intensity, S_f is a slope factor, and C_i is a cover factor.

The diameter, density, and shape of sediment particles determine their transportability and their deposition. Sediment eroded from most cultivated soils is a mixture of primary particles and aggregates (*24*). An important development in erosion prediction was to consider the characteristics of the sediment as it was detached, rather than dispersing the sediment in laboratory tests and using the characteristics of the primary particles of the dispersed sediment to describe its transport and deposition characteristics (*28*).

Only in the past decade have researchers recognized the significant contribution of ephemeral gully erosion to the total sediment production and the role of ephemeral gullies in delivering sediment from fields. An ephemeral gully differs from a classical gully in that it is small enough to be removed by normal tillage. In some situations, sediment production by ephemeral gully erosion can equal or exceed that produced by rill and interrill erosion (*40*). In estimating ephemeral gully erosion (*16, 41*), the gully is assumed to erode vertically downward until a less erodible layer is reached. The gully then widens until the depth of flow is so shallow that detachment and/or transport of additional sediment is no longer possible (Figure 2). The equations used to model the ephemeral gully process are similar to those used for rill erosion (Equation 7).

Ephemeral gully erosion reshapes the topography. Annual tillage operations tend to fill the gullies eroded during the previous year. This soil movement and incisement of a gully steepens the areas adjacent to the gully, which, in turn, accelerates rill and interrill erosion on adjacent areas. When ephemeral erosion rates are high, the amount of reshaping becomes significant over time. At present, the change in landscapes with erosion or deposition is not considered; landscape shape is assumed to be static in the analysis. However, process-based erosion prediction technically has the potential for describing this landscape evolution.

With FSA-associated use of erosion prediction in policy development and implementation, the limitations of empirical erosion prediction technology became apparent. At the time, however, adequate personal computers and access to soil, cropping, and weather data bases were becoming generally available to research scientists and field staff. Experience gained with erosion simulation models and the accumulation of research findings over the last two decades on fundamental erosion processes made the development of a process-based model for general use a possibility.

Against this background, several agencies, including ARS, SCS, Forest Service, Bureau of Land Management, U.S. Geological Survey, and other cooperators initiated the Water Erosion Prediction Project (WEPP); this project is aimed at developing a new generation, process-based prediction technology to replace the USLE (*13*).

A similar project is also underway to develop process-based erosion prediction technology to replace the WEQ with a computer model called Wind Erosion Prediction System (WEPS) (*20*).

A special feature of these prediction projects is that the experimental research has been conducted across the United States to determine soil erodibility for detachment by raindrop impact separate from soil erodibility for detachment by flow (*9, 35*). Also the effects of cropping and management on the separate processes are being studied (*1*). Applications of the products from the water and wind erosion prediction projects will begin to be used in local SCS offices in the mid-1990s.

Challenges in Erosion Prediction Research

Although WEPP and WEPS will produce major advancements in erosion prediction technology, several important scientific challenges remain. One of these challenges is to perfect the approach of applying, in a practical sense, the force-resistance concept to determining erodibility. Even in the most process-based models, erodibility is not a soil property; it results, by definition, from an erosivity term and is evaluated by experiment. It would be preferable to view erodibility by considering the local and instantaneous forces applied to the soil by raindrop impact and the internal forces in the soil as the soil responds to the externally applied forces. Science is not yet able to sufficiently measure and predict both the external and internal forces. The fall cone penetrometer and classical soil mechanics tests are being studied as ways to measure these forces (*2, 8, 10*), but these methods may not be measuring the same forces on the same scale as experienced by soil as it fails when detachment occurs.

Another major challenge is to deal with local variability in the erosive forces. A scheme is needed to integrate these forces over time and over space into a practical working variable that can be used in erosion prediction. For example, in the case of raindrop impact detachment, erosivity should eventually be related to rainfall intensity because this variable is the standard input for rainfall. A much greater problem is to integrate the resistance forces in the soil over time and space. For example, the strength of a clod could be much greater than the strength of particles between clods. A practical method is needed to "measure" erodibility easily and quickly in a field setting. Much of the erodibility research has been conducted on idealized or standard soil conditions. This research must give more consideration to how erodibility changes with time in field situations as it is influenced by climate, cropping, tillage, and management.

In many situations where deposition occurs, the sediment leaving an area is determined by the transport capacity of the surface runoff. Most relationships now used to describe sediment transport capacity were derived for flow and sediment conditions typical of rivers, rather than flow and sediment conditions typical of farm fields (*12, 19*). A particularly critical need is to describe the transport capacity of thin flow on interrill areas. A concentrated research effort is needed to develop new theory and to collect data for validation of the new theory (*19*).

Erosion modeling for water erosion generally is based on the assumption of broad sheet flow (*12*). Such flow often does not occur in fields, however. The assumption of sheet flow, regardless of the rigor of the theory or mathematics, greatly distorts parameters from values determined in careful laboratory experiments. In fact, equations based on the broad sheet flow assumption can never describe the processes as they occur in the field, regardless of how well the theory works for idealized laboratory conditions. One approach to this problem is the stochastic generation of flow and rill patterns, analysis of selected rills, and aggregation of results to estimate erosion for a large area.

Scale is a major problem in erosion prediction. As a process, erosion occurs on a particle scale, but the particular interest is in the field and landscape scales. The challenge is to be able to compute erosion on the particle scale and to integrate the result in time and space to a much larger scale of interest, while minimizing the number of computations feasible for practical application of the erosion prediction technology.

Detachment and deposition change soil depth. Moreover, these poorly understood processes are selective where fine particles are removed and coarse ones are left. How erosion changes soil properties is not completely understood and certainly is not modeled in a sophisticated way in any current erosion model. Therefore, study of erosion under steady conditions may reveal little about the mechanism of selectivity, except for special conditions where deposition is being induced, as in depressional areas or low grade furrows.

Summary

Erosion prediction research began in the 1930s. By the 1960s empirical methods of estimating erosion were well established. With the development and availability of personal computers, a new technology of process-based erosion prediction is possible, and numerous models have been and are continuing to be developed to replace the empirical models. Research on many of the processes is still needed, particularly those that describe overland flow and spatial and temporal variation in a soil's susceptibility to erosion.

REFERENCES CITED

1. Alberts, E. E., M. A. Weltz, and F. Ghidey. 1989. *Plant growth component.* In L.S. Lane and M.A. Nearing [editors] *USDA-Water Erosion Prediction Project; Hillslope Profile Model Documentation.* NSERL Report No. 2. National Soil Erosion Research Laboratory, Agricultural Research Service, U.S. Department of Agriculture, West Lafayette, Indiana.
2. Al-Durrah, M., and J. M. Bradford. 1981. *New methods of studying soil detachment due to waterdrop impact.* Soil Science Society of America Journal 45: 949-953
3. Beasley, D. B., L. F. Huggins, and E. J. Monke. 1980. *ANSWERS: A model for watershed planning.* Transactions, American Society of Agricultural Engineers 23(4): 938-980.
4. Bone, S. W., R. Christman, L. M. Feusner, R. L. Goettemoeller, and B. H. Nolte. 1979. *Ohio erosion control and sediment pollution abatement guide.* Cooperative Extension Service, Ohio State University, Columbus.
5. Chepil, W. S. 1959. *Wind erodibility of farm fields.* Journal of Soil and Water Conservation 14: 214-219.
6. Cole, G. W. 1984. *A method for determining field wind erosion rates from wind-tunnel-derived functions.* Transactions, Americian Society of Agricultural Engineers 27(1): 110-116.
7. Cook, L. H. 1936. *The nature and controlling variables of the water erosion process.* Soil Science Society of America Proceedings 1: 487-494.
8. Cruse, R. M., and E. W. Larson. 1977. *Effect of soil shear on soil detachment due to raindrop impact.* Soil

Science Society of America Journal 41: 777-781.

9. Elliot, W. J., A. M. Liebenow, J. M. Laflen, and K. D. Kohl. 1989. *A compendium of soil erodibility data from WEPP cropland soil field erodibility experiments 1987 and 88.* NSERL Report No. 3. National Soil Erosion Research Laboratory, Agricultural Research Service, U.S. Department of Agriculture, West Lafayette, Indiana.
10. Elliot, W. J., L. J. Olivieri, J. M. Laflen, and K. D. Kohl. 1990. *Predicting soil erodibility from soil strength measurements.* Paper No. 90-2009, American Society of Agricultural Engineers, St. Joseph, Michigan.
11. Ellison, W. D. 1947. *Soil erosion studies—part I.* Agricultural Engineering 28(4): 145-146.
12. Finkner, S. C., M. A. Nearing, G. R. Foster, and J. E. Gilley. 1989. *A simplified equation for modeling sediment transport capacity.* Transactions, American Society of Agricultural Engineers 32(5): 1,545-1,550.
13. Foster, G. R. 1987. *User requirements, USDA-water erosion prediction project (WEPP).* Draft 6.2. National Soil Erosion Laboratory, Agricultural Research Service, U.S. Department of Agriculture, West Lafayette, Indiana.
14 Foster, G. R. 1982. *Modeling the erosion process.* In C. T. Haan, H. P. Johnson, and D. L. Brakensiek [editors] *Hydrologic Modeling of Small Watersheds.* Monograph No. 5. American Society of Agricultural Engineers, St. Joseph, Michigan.
15. Foster, G. R., L. D. Meyer. 1972. *Mathematical simulation of upland erosion by fundamental erosion mechanics.* In proceedings, Sediment-Yield Workshop. U.S. Department of Agriculture, Oxford, Missippi.
16. Foster, G. R., L. J. Lane, J. Nowlin, J. M. Laflen, and R. A. Young. 1980. *A model to estimate sediment yield from field-size areas: Development of model.* In W. G. Knisel [editor] *CREAMS: A Field Scale Model for Chemicals, Runoff, and Erosion from Agricultural Management Systems.* Conservation Research Report No. 26. U. S. Department of Agriculture, Washington, D. C. pp. 36-64.
17. Foster, G. R., and R. E. Highfill. 1983. *Effect of terraces on soil loss: USLE P factor values for terraces.* Journal of Soil and Water Conservation 38(1): 48-51
18. Foster, G. R., and W. H. Wischmeier. 1974. *Evaluating irregular slopes for soil loss prediction.* Transactions, American Society of Agricultural Engineers 17(2): 305-309.
19. Guy, B. T., W. T. Dickinson, and R. P. Rudra. 1990. *Comparison of fluvial transport equations as applied to shallow overland flow.* Paper No. 90-2014, American Society of Agricultural Engineers, St. Joseph, Michigan.
20. Hagen, L. J. 1988. *Wind erosion mechanics: Abrasion of aggregated soil.* Paper No. 88-2561. American Society Agricultural Engineers, St. Joseph, Michigan.
21. Knisel, W. G. 1980. *CREAMS: A field scale model for chemicals, runoff, and erosion from agricultural management systems.* Conservation Research Report No. 26. Agricultural Research Service, U.S. Department of Agriculture, Washington, D.C. 640 pp.
22. Laflen, J. M., G. R. Foster, and C. A. Onstad. 1985. *Simulation of individual storm loss for modeling the impact of soil erosion on crop productivity.* In S. A. El-Swaify, W. C. Moldenhaur, and Andrew Low [editors] *Soil Erosion and Conservation.* Soil Conservation Society of America, Ankeny, Iowa. pp. 285-295.
23. Lane, L. J., and M. A. Nearing. 1989. *USDA-Water erosion prediction project: Hillslope profile model documentation.* NSERL Report No. 2. National Soil Erosion Research Laboratory, Agricultural Research Service, U.S. Department of Agriculture, West Lafayette, Indiana.
24. Line, D. E., and L. D. Meyer. 1990. *Transport of sediment from soils of different textures.* Paper No. 90-2015. American Society of Agricultural Engineers, St. Joseph, Michigan.
25. McCool, D. K., G. R. Foster, C. K. Mutchler, and L. D. Meyer. 1989. *Revised slope length factor for the universal soil loss equation.* Transactions, American Society of Agricultural Engineers 32(5): 1,571-1,576.
26. McCool, D. K., L. C. Brown, G. R. Foster, C. K. Mutchler, and L. D. Meyer. 1987. *Revised slope steepness factor for the universal soil loss equation.* Transactions, American Society of Agricultural Engineers 30(5): 1,387-1,396.
27. Meyer, L. D., D. G. DeCoursey, and M.J.M. Romkens. 1977. *Soil erosion concepts and misconceptions.* In G. R. Foster [editor] *Soil Erosion: Prediction and Control.* Soil Conservation Society of America, Ankeny, Iowa. pp. 2-12.
28. Meyer, L. D., and S. H. Scott. 1983. *Possible errors during field evaluations of sediment size distributions.* Transactions, American Society of Agricultural Engineers 26(2): 481-485, 490.
29. Meyer, L. D., and W. H. Wischmeier. 1969. *Mathematical simulation of the process of soil erosion by water.* Transactions, American Society of Agricultural Engineers 12(6): 754-758, 762.
30. Musgrave, G. W. 1947. *The quantitative evaluation of factors in water erosion, a first approximation.* Journal of Soil and Water Conservation 2(3): 133-138.
31. Negev, M. 1967. *A sediment model on a digital computer.* Technical Report 76. Department of Civil Engineering, Stanford University, Stanford, California. 109 pp.
32. Renard, K. G., and J. R. Simanton. 1990. *Application of RUSLE to rangelands.* In Proceedings of Symposium on Watershed Management. American Society of Civil Engineers, New York, New York.
33. Robinson, A. R. 1971. *Sediment: Our greatest pollutant?* Agricultural Engineering 52(8): 407-408.
34. Sheridan, J. M., C. V. Booram, Jr., L. E. Asmusen. 1982. *Sediment-delivery ratios for a small coastal plain agricultural watershed.* Transactions, American Society Agricultural Engineers 25(3): 610-615, 622.
35. Simanton, J. R., L. T. West, M. A. Weltz, and G. D. Windgate. 1987. *Rangeland experiments for the water erosion prediction project.* Paper No. 87-2545. American Society of Agricultural Engineers, St. Joseph, Michigan.
36. Smith, D. D., and D. M. Whitt. 1947. *Estimating soil losses from field ares of claypan soils.* Proceedings, Soil Science Society of America 12: 485-495.
37. Soil Conservation Service. 1985. *Technical guide section I C 1.* Des Moines, Iowa. pp. 15 19.
38. Soil Conservation Service. 1988. *Part 502—wind erosion.* National Agronomy Manual. U.S. Department of Agriculture, Washington, D.C.
39. Stallings, J. H. 1950. *Erosion of topsoil reduces productivity.* SCS-TP-98. U.S. Department of Agriculture, Washington, D.C.
40. Thomas, A. W., R. Welch, and T. R. Jordan. 1986. *Quantifying concentrated-flow erosion on cropland with aerial photogrammetry.* Journal of Soil and Water Con-

servation 41(4): 249-252.

41. Watson, D. A., J. M. Laflen, and T. G. Franti. 1987. *Estimating ephemeral gully erosion.* National Soil Erosion Research Laboratory, Agricultural Research Service, U.S. Department of Agriculture, West Lafayette, Indiana.
42. Williams, J. R. 1975. *Sediment-yield prediction with universal equation using runoff energy factor.* In *Present and Prospective Technology for Predicting Sediment Yields and Sources.* ARS-540. U.S. Department of Agriculture, Washington, D.C. pp. 244-251.
43. Williams, J. R., C. A. Jones, J. R. Kiniry, and D. A. Spanel. 1989. *The EPIC crop growth model.* Transactions, American Society of Agricultural Engineers 32(2): 497-511.
44. Wilson, B. N., B. J. Barfield, A. D. Ward, and I. D. Moore. 1984. *A hydrology and sedimentology watershed model. Part I: Operational format and hydrologic component.* Transactions, American Society of Agricultural Engineers 27(5): 1,370-1,377.
45. Wischmeier, W. H. 1976. *Use and misuse of the universal soil loss equation.* Journal of Soil and Water Conservation 31(1): 5-9.
46. Wischmeier, W. H., and D. D. Smith. 1960. *A universal soil loss estimating equation to guide conservation farm planning.* In Transactions, Seventh Congress. International Soil Science Society, Wageningen, The Netherlands. pp. 418-425.
47. Wischmeier, W. H., and D. D. Smith. 1965. *Predicting rainfall-erosion losses from cropland east of the Rocky Mountains.* Agriculture Handbook 282. U.S. Department of Agriculture, Washington, D.C.
48. Wischmeier, W. H., and D. D. Smith. 1978. *Predicting rainfall erosion losses—a guide to conservation planning.* Agriculture Handbook 537. U.S. Department of Agriculture, Washington, D.C.
49. Woodruff, N. P., and F. H. Siddoway. 1965. *A wind erosion equation.* Soil Science Society of America Proceedings 29: 602-608.
50. Young, R. A., C. A. Onstad, D. D. Bosch, and W. P. Anderson. 1987. *AGNPS, Agricultural nonpoint-source pollution model. A watershed analysis tool.* Conservation Research Report 35. U.S. Department of Agriculture, Washington, D.C. 80 pp.
51. Zingg, R. W. 1940. *Degree and length of land slope as it affects soil loss in runoff.* Agricultural Engineering 21: 59-64.

6

Erosion Productivity Impact Prediction

F. J. Pierce

Soil erosion affects the productivity of a soil in significant ways. Through soil removal or sediment deposition, erosion alters the inherent physical and chemical properties of soils. These alterations, in turn, affect the processes that regulate the productivity of an ecosystem.

Although erosion-induced changes in soil can lead to improved productivity, as in the case of some depositional environments or where erosion exposes soils with more favorable properties, erosion represents the major agent of soil degradation worldwide (*34, 63*). In addition to soil degradation, erosion also damages plants during their life cycles, resulting in lost productivity. The nature and extent of these damages are difficult to quantify, however.

Even though the cumulative body of literature is extensive (*3, 8, 13, 14, 15, 17, 28, 34, 39, 40, 41, 45, 46, 47, 48, 61, 62, 64, 65, 67, 68, 91, 108, 109, 121*), the large-scale effects of erosion on the productivity of soils are not well known. Lal (*62*) concluded that, in spite of billions of dollars invested in the erosion problem, "we cannot say for sure what effect the loss of a unit of soil depth has on crop yield."

The evidence from erosion studies has led to certain generally accepted conclusions. Most notable are that soil erosion has been accelerated over natural, geologic rates by human activities and that, left uncontrolled, soil erosion is a major agent in the degradation of soil worldwide. A third and important conclusion is that soil erosion and its impact on soil productivity are highly variable from place to place and over time. It is the different scales of space and time that explain, for example, why over broad areas erosion has increased while at the same time productivity of soils has increased. This also explains why soils once considered permanently lost to erosion damage have not proven to be so (*131*). The major concerns about erosion are (a) the rate and extent to which erosion is degrading soils worldwide and the extent to which this degradation will limit food and fiber production for a growing world population and (b) the off-farm impacts of soil erosion, for example, nonpoint-source pollution of surface water resources (*24*).

Much of the recent interest in soil erosion relates to prospects for tomorrow (*131*), and efforts in the 1980s were directed toward the development of predictive capabilities. The need for predictive capabilities was summarized by Wolman (*131*) as follows:

"Global assessment is important in planning for agricultural development, in assessing the adequacy of food resources for the world's population and in evaluating agricultural policies.... Undoubtedly, the development of ...

[such] models will improve, or refine, estimates of the aggregate potential impacts of erosion on a world scale ... and provide improved techniques for evaluating alternatives and allocating scarce resources where they can be used most effectively and efficiently at the local, county, or regional level.''

It was recognized early in the development of predictive technologies that attempts to quantify erosion's effects on productivity were generally inadequate for this purpose (*128*). This is not to be taken as overly critical of erosion research to date, but rather to recognize its site-specific and short-term nature. Studies that have documented productivity losses from soil erosion have been limited to relatively few soils and have been conducted under specific weather, cropping, and management conditions. The effects of erosion on loss of productivity in any given year can be small relative to changes associated with annual variations in climate. Also, long-term changes in productivity often are difficult to establish because of increased productivity from improved technology. Therefore, extrapolating results from site-specific studies to regional and global scales is only possible with the development of predictive capabilities in conjunction with extensive geographically referenced data.

The Complexity of the Soil Erosion-Productivity Relationship

A prerequisite to predicting the impacts of erosion on soil productivity is understanding the nature of the soil erosion-productivity relationship. Experiments to measure erosion's effect have been the primary tool for studying effects of past erosion on productivity. Most of these studies have attempted to determine the effect of past erosion on current productivity levels and, concurrently, to determine the relative importance of soil variables to measured productivity (*62, 72*). These relationships have been used to quantify the erosion-productivity relationship for soils as affected by erosion's impact on inherent soil properties and related processes.

A graphic example of an erosion-productivity relationship for an eroding soil is illustrated in figure 1 (redrawn from *99*). This graph illustrates the average trend line in the relationship between soil productivity and time, eliminating annual fluctuations due, for example, to weather and spatial variability for a given soil. While experimental data describe portions of this relationship for given soils for specific weather, crop, and management conditions (and under certain underlying assumptions), the relationship, as depicted, is not really known for any soil. However, knowledge of this relationship for spatially distributed soils is needed for resource assessment and has been the goal of prediction technology.

Prediction of Past Erosion's Effects on Productivity

Knowledge about past productivity levels of soils or their change with time in relation to soil development and soil degradation processes is limited. Therefore, the level of soil productivity and its rate of change prior to accelerated erosion must be surmised from present productivity levels. This approach has two inherent problems.

First, soil development and degradation processes determine the intrinsic properties of a soil, while productivity is determined by the efficiency in the use and management of resource inputs, including soil, water, energy, climate, fertilizers, pesticides, labor, and machinery (*130*). As managed inputs increase or agriculture becomes more technically advanced, the relative contribution of the soil to crop yield diminishes. Therefore, as Blaikie (*10*) suggests, ''The principal issue is a question of judgement over the extent to which land-users can mask or 'make good' soil degradation and erosion, and this depends upon their present and future access to chemical fertilizers, improved seeds, credit, government assistance in land erosion works, soil conserving crop rotations and tillage methods.'' Present levels of crop yields, therefore, better reflect crop productivity as defined by Wittwer (*30*) than soil productivity in a geologic sense.

Second, estimates of past productivity require knowledge of past erosion and associated changes in productivity. Because erosion is

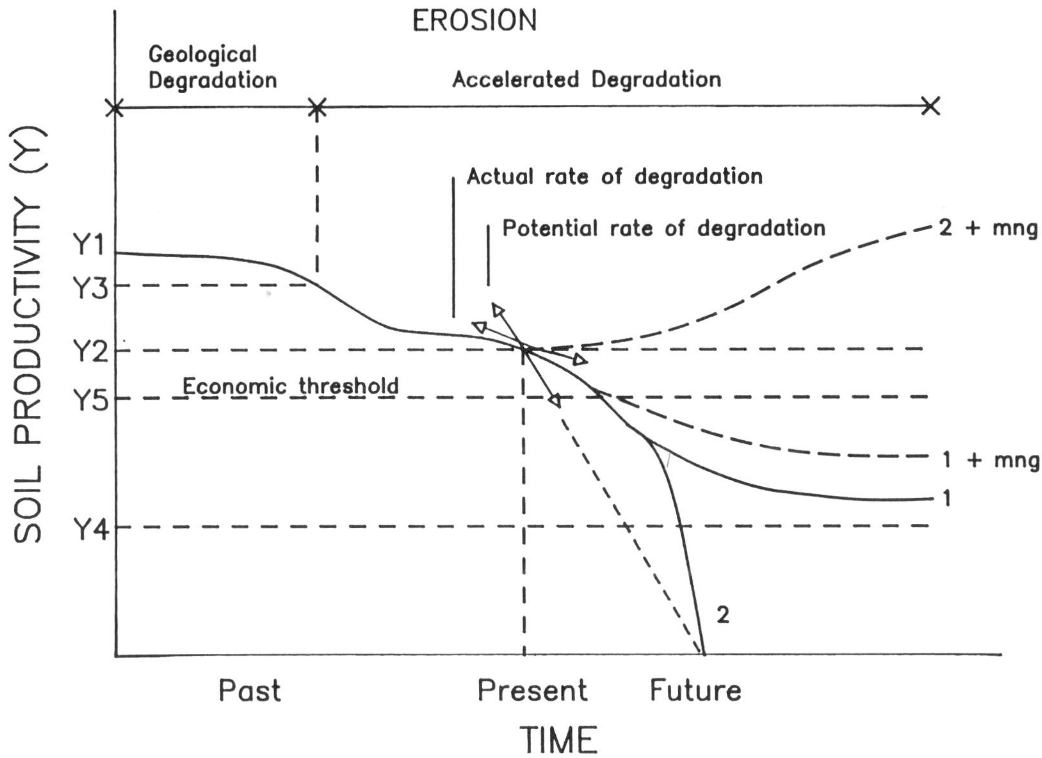

Figure 1. Changes in soil productivity as influenced by the degradation < - > aggradation processes. Y1, Y2, and Y3 represent soil productivity initially, at present, and at the start of accelerated degradation, respectively. Curve 1 represents the case where cultivation is practiced for several years without fertilization, thus leading to asymptotic low productivity level, Y4. Curve 2 represents accelerated degradation ending with complete elimination of soil productivity (e.g., complete loss of a shallow soil). Curve 1-mng depicts improved management over curve 1, whereby erosion is stabilized after an initial decrease in soil productivity. Curve 2-mng represents successful management resulting in rebuilding of soil productivity. (Redrawn from *40*, after *99*).

spatially variable and confounded with landscape position, it is difficult to attribute differences in yield within a field to erosion-induced changes or to changes associated with different rates of soil formation regulated by landscape position (*23, 29, 30, 31, 32*). A popular experimental approach has been to measure soil properties and relate them to crop yield. This has been done for soils in-situ or in disturbed systems, such as reclamation, artificial soil removal or addition, and in greenhouse studies. The problem with this approach is that the soil properties that regulate soil productivity must be identified and quantified. Even if this were possible, soil properties identified as highly correlated with crop yields often are related to both erosion and soil genesis on a landscape position basis (*23*). Therefore, the level of soil productivity prior to the initiation of accelerated erosion (Y2 in Figure 1) is difficult to know.

Erosion research in the 1930s and 1940s (Table 1) established an inverse relationship between yield and topsoil depth. This early work was thought to provide sufficient evidence of the detrimental effects of soil erosion on crop yields, so this aspect of erosion research was phased out in the early 1950s. Emphasis was then given to the causes and rates of erosion and techniques for controlling it (*123*). Research from the 1950s through 1970s (Table 1) continued to add to the evidence about the detrimental effects of erosion.

More recently, U.S. legislation—the Soil and Water Resources Conservation Act (RCA) of 1977—directed the secretary of agriculture

Table 1. Summary of selected studies relating to the soil productivity effects of erosion.

Soil Series	*Location*	*Type of Study**	*Years of Study*	*Crop*	*Notes*	*Author*
Amarillo	TX	PE	1962	millet	Topsoil outyielded subsoil when no fertilizer or N alone was applied. Subsoils outyielded topsoils under P alone and N and P treatments.	Eck and Ford, 1963
Austin	TX	EE	1961-1965	sorghum	Yields were lower on desurfaced plots where bulk densities were highest near the surface.	Smith et al., 1967
Beadle	SD	ER	1966-1973	corn	Corn yields were not restored with fertilizer applications.	Olson, 1977
Bladen	NC	PE	—	wheat	Yields were less on cut than on the unaltered soil even with addition of P, K and lime.	Phillips and Kamprath, 1973
Brookston Parr Fowler Crosby Miami	IN	EE	—	corn	Yield ranged in 1939 from 49 bu/ac for locations with 1 to 2 in. topsoil to 93 bu/ac for those having 13 or in. of topsoil.	Uhland, 1949
Canfield	OH	EE	1940-1956	corn	Yields were considerably lower on desurfaced plots. These differences were reduced after 12 years of crop rotations (see Figure 1).	Bachtell et al., 1956
Carrington	MN	PE	1939	clover	Subsoils similar yields similar to surface soils in most horizons.	Rost, 1940
Cecil	GA	EE	1944-1948	corn cotton oats vetch	Reduction in yield per in. soil per acre was 2.7 bu for corn, 104 lbs. for cotton, 3.8 bu for oats, and 560 lbs. for vetch	Adams, 1949
	SC	EE	1936-1939	cotton	Field study of A, B, and C horizons of this soil. Yield ratios of A/B and A/C were 3 and 11. All cotton yields were <1000 lbs/ac.	Latham, 1940
Celina	OH	EE	1940-1956	corn	Yields were considerably lower on desurfaced plots. These differences were reduced after 12 years of crop rotations.	Bachtell et al., 1956
Collington	NJ	EE	1938-1941	potato	Yield related to depth of topsoil.	Neal, 1943
Congaree	NC	PE	—	wheat	Soil from cut area was more productive than soil from unaltered site.	Phillips and Kamprath, 1973
Crider	KY	EE	1977-1980	corn	Corn yield averaged 21% less on plots over the years of study due primarily to reduced water availability.	Frye et al., 1982
Dolhart	TX	PE	1962	millet	Topsoil outyielded subsoil when no fertilizer or N alone was applied. Subsoils outyielded topsoils under P alone and N and P treatments.	Eck and Ford, 1962

Continued

Table 1. Summary of selected studies relating to the soil productivity effects of erosion *(continued).*

Soil Series	*Location*	*Type of Study**	*Years of Study*	*Crop*	*Notes*	*Author*
Dubuque	MN	PE	1939	clover	Subsoils similar yields similar to surface soils in most horizons. Deeper horizons less productive.	Rost, 1940
Fayette	WI	EE	1939-1947	corn	Yields on severely eroded sites increased over time to equal that of moderately eroded sites.	Hays et al., 1948
	MN	PE	1939	clover	Under fertilization, yields on subsoils were similar to surface soils	Rost, 1940
Fort Collins	CO	LL	1946-1947	Var.	Fertilization with N and P restored productivity to leveled land.	Whitney et al., 1950
Gardena	ND	PE	1957	corn	Yield on Ap 1 horizon $>$AC horizon$=$C horizon. The pH of C horizon $>$8.0.	Carlson and Grunes, 1958
	ND	LL	1954-1956	corn	Irrigated. Subsoil yields were excellent when a combination of N, P and manure were added.	Carlson et al., 1961
Gloucester	MN	PE	1939	clover	Subsoils similar yields similar to surface soils in most horizons.	Rost, 1940
Goldsboro	NC	PE	—	wheat	Yields were less on cut than on the unaltered soil even with addition of P, K and lime.	Phillips and Kamprath, 1973
	NC	PE	—	wheat	Yields were less on cut than on the unaltered soil due to poor subsoil conditions.	Thomas and Cassel, 1979
Groseclose	VA	EE	1965-1968	corn	Yields on irrigated, mulched + irrigated, and mulched subsoil were equivalent to irrigated topsoil under fertilization.	Batchfield and Jones, 1972
Hempstead	MN	PE	1939	clover	Subsoils similar yields similar to surface soils in most horizons.	Rost, 1940
Hildago	TX	LL	1959	sorghum	Irrigated. Forage yield slightly less on cut areas than on unaltered.	Heilman and Thomas, 1961
	TX	LL	1972	sorghum	Irrigated. Forage dry matter affected by land leveling method.	Thomas et al., 1974
Keene Muskingum Coshocton	OH	EE	—	corn	Average yields on 3 watersheds ranged from 33.7 bu/ac with 4 in. of topsoil to 59.5 bu/ac with 9 in. of topsoil.	Uhland, 1949
Keiser	MT	LL/PE	—	corn	N and P fertilization restored yields on subsoil plots to those of surface soil.	Reuss and Campbell, 1961
Lindley	MN	PE	1939	clover	Subsoils similar yields similar to surface soils in most horizons.	Rost, 1940

Continued

Table 1. Summary of selected studies relating to the soil productivity effects of erosion *(continued)*.

Soil Series	*Location*	*Type of Study**	*Years of Study*	*Crop*	*Notes*	*Author*
Lynchburg	NC	PE	—	wheat	Yields were less oncut than on the unaltered soil even with addition of P, K and lime.	Phillips and Kamprath, 1973
Lynchburg	NC	PE	—	wheat	Yields were less on cut than on the unaltered soil due to poor subsoil conditions.	Thomas and Cassel, 1979
Marshall	IA	EE	1958-1959	corn	Equal corn yields obtained on surface and subsoil under adequate fertility.	Langdale and Shrader, 1961
	IA	EE	1957-1958	corn	Under fertilization, subsoil yields were equivalent in 1957 and less than the surface soil in 1958.	Englestead et al., 1961
I	IA	EE	1939	corn	Yield reduction was related to the depth of topsoil, 120 bu/ac (12 in) and 52 bu/ac (1-2 in).	Uhland, 1949
Maury	KY	EE	1977-1980	corn	Corn yield averaged 12% less on plots over the years of study due primarily to reduced water availability.	Frye et al., 1982
Miles	TX	PE	1962	millet	Topsoil outyielded subsoil when no fertilizer or N alone was applied. Subsoils out-yielded topsoils under P alone and N and P treatments.	Eck and Ford, 1962
Monona	IA	EE	1957-1958	corn	Under fertilization, subsoil yields were equivalent in 1957 and less than the surface soil in 1958.	Englestead et al., 1961
Muscatine	IL	MR	1975-1977	corn	Topsoil over mine spoil returned yields to that of the Muscatine. Graded spoil alone yielded only half of Muscatine soil.	Grandt, 1978
Newtonia	AK	EE	1935-1953	corn	Yield maximum 64 bu/ac. Large yield reductions followed topsoil removal. Drought conditions occurred 6 of 19 years.	Bartholomew et al., 1954
Norfolk	NC	PE	—	wheat	Crop growth reduced in cut areas.	Phillips and Kamprath, 1973
Nunn	MT	LL/PE	—	corn	N and P fertilization restored yields on subsoil plots to those of surface soil.	Reuss and Campbell, 1961
Portsmouth	NC	PE	—	wheat	Crop growth reduced in cut areas.	Phillips and Kamprath, 1773
Pullman	TX	LL	1960-1962	sorghum	Irrigated 0-41 cm soil removed. Similar yield potentials obtained on all but the 41 cm removed plots.	Eck et al., 1965
	TX	LL	1963-1965	sorghum	Pre-irrigation only. Yields on cut plots <undisturbed plots. Differences attributed to water storage and use.	Eck, 1969

Continued

Table 1. Summary of selected studies relating to the soil productivity effects of erosion *(continued)*.

Soil Series	*Location*	*Type of Study**	*Years of Study*	*Crop*	*Notes*	*Author*
Pullman	TX	PE	1963	millet	With fertilization, only the 13-19 in. soil depth produced yields lower than the surface soil	Eck et al., 1965
Rains	NC	PE	—	wheat	Yields were less on cut than on the unaltered soil even with addition of P, K and lime.	Phillips and Kamprath, 1973
	NC	PE	—	wheat	Yields were less on cut than on the unaltered soil due to poor subsoil conditions.	Thomas and Cassel, 1979
Shelby Grundy	MO	EE	—	corn	Yields reduced with topsoil depth up to 90% in a given year.	
Swygert	IL	EE	1946-1947	corn	Yield related to topsoil depth (2.7 bu/in-ac).	Odell, 1950
Tama	IL	EE	1946-1947	corn	Yield related to topsoil thickness. (1.2 bu/in-ac)	Odell, 1950
	MN	PE	1939	clover	Yield of subsoil similar to topsoil.	Rost, 1940
	IA	EE	1939	corn	Yields related to topsoil thickness, 64 bu/ac (12 in) and 32 bu/ac (1.5 in).	Uhland, 1949
Temvik	ND	MR	1975-1979	wheat	Subsoils similar yields to surface soils in most horizons. Deeper horizons slightly less.	Power et al., 1981
Wagman	NC	PE	—	wheat	No difference in plant growth in on unaltered or cut soil with addition of P and K.	Phillips and Kamprath, 1973
Weld	CO	EE	1957-1961	wheat	Wheat-fallow system, 0-38.1 cm of soil removed. Yield differences only on plots with deepest soil removed.	Black and Greb, 1968

*Five groups: EE - those dealing directly with erosion's effects, LL - land-leveling, MR - mining reclamation, ER - erosion restoration, and PE - greenhouse or pot-experiments.

to continually appraise the status and condition of and trends in soil, water, and related natural resources in the United States beginning in 1980 and every five years thereafter. This stimulated considerable research effort to supplement the limited data available to document erosion's effects on soil productivity (*128*). Much of this research was reported in three major symposiums (*3, 41, 45*).

The results from studies over the last 50 years have varied. What follows are nine general observations that can be summarized from these studies. Again, it must be emphasized that many of these studies are confounded by scales of space and time.

1. Yield levels reported in many of these studies were low relative to present production levels and much of the data were for a few years only. While many studies have related crop yields to soil properties, there is no agreement as to which soil properties regulate soil productivity and there has been no standardization in their measurement. These difficulties prohibit the establishment of long-term trends and make it difficult to assess the effects of improved technology on the erosion-productivity relationship (*86*). In addition, the contribution of soil to crop yield as a percentage of yield has declined. Crop yields are often greater on soils under properly managed

inputs than in their virgin condition (*58, 60*). Krantz (*60*) may have said it best, "The present day farmer is dependent far less upon inherent fertility and other soil properties as he has much more opportunity to modify the soils to fit his needs more nearly."

2. It was recognized early that in many cases management inputs were sufficient to restore production to levels of undisturbed soils and that the degree to which this was possible related to the characteristics of the subsoil (Table 1). High mechanical impedance, high aluminum saturation, low pH, reduced plant-available water-holding capacity, increased clay content, and a reduced root zone due to the presence of bedrock were cited by Winters and Simonson (*129*) in 1952 as factors limiting yield beyond that compensated by the addition of fertilizers.

3. Under limited or no fertilization, yields often were highly related to depth of topsoil. The inverse nature of the relationship between yield and topsoil depth allowed the calculation of the rate of yield loss per unit depth of soil loss (*2, 68*). Use of these rates (*115*) led to an early conclusion that productivity reductions were proportional to the erosion rate, implying soil would be removed from production when the topsoil was eroded. This notion has been used as one method of estimating soil productivity loss (*75*). However, topsoil thickness is spatially related to landscape position and to erosion history.

4. There is a relationship between crop yield and soil depth. Thomas and associates (*118*) described effective soil depth as an important factor in crop yield because of its role as the major determinate of water storage capacity of soil, while topsoil depth determines productive capacity because it contains the plant nutrients vital for plant growth. Both soil depth and topsoil depth have been used as criteria in mined-land reclamation (*92*). However, the depth of topsoil needed for reclamation depends on the characteristics of the subsoil material (*85, 105*). It is also clear that results can vary with year (*42, 49*).

5. The permanency of losses in production from erosion described in these studies is uncertain. Bachtell and associates (*4*) demonstrated for artificial soil removal on the Canfield and Celina soils in Iowa that corn yields recovered lost productivity in time, depending on soil type, management inputs, and crop rotation. Hays and associates (*54*) reported similar results for the Fayette soil in Wisconsin. Lal (*62*) states that 40 million hectares of land considered by Bennett (*8*) to be ruined or severely damaged by erosion by the mid-1930s is now producing profitable returns with the development of improved technologies.

6. The ability to find uneroded sites may be limited. Thomas and associates (*119*) concluded that the magnitude of the differences between eroded and uneroded sites may be difficult to assess because of the scarcity of uneroded sites in upland Piedmont soils in southern Virginia. Therefore, lack of standards for comparison as to the initial level of productivity may preclude the determination of historic productivity losses from soil erosion.

7. Other effects of erosion on soil productivity have been largely ignored. These include effects of gully and ephemeral erosion, landslides, and the associated loss of contiguous parcels of land. They include effects associated with sediment deposition, including crop damage, which, along with productivity changes, has important short-term effects on subsistence farmers in developing countries (*40*). Productivity changes can be positive (soil-building) or negative (degrading) or both within the same field unit. Carter and associates (*22*) showed that in furrow-irrigated lands in Idaho erosion redistributed topsoil resulting in a continuous topsoil depth gradient of 10 to 66 centimeters in fields. Yields of all crops studied increased with increasing topsoil depth. However, yield decreases per unit of topsoil removal were larger than per unit depth increases resulting from deposition. Erosion-induced changes in soil properties also affect regional hydrology, resulting in crop losses because of increased flooding in low areas (*40*) or losses associated with increased leaching of nutrients resulting from the absence in degraded soils of plant roots, which normally recycle nutrients within the soil profile (*5, 40*).

8. The effects of erosion on soil productivity often are hard to visualize. They are also long-

term and, at least temporarily, often masked by technology (*62*). The soil system may be buffered in such a way that the erosion-induced changes in soil properties regulating soil productivity have not exceeded critical limits for soil degradation. Thus, there may be considerable time after accelerated erosion begins before soil removal or sediment deposition has any measurable impact on crop yield. In addition, the effects on productivity of soil removal by erosion can be positive for certain soil conditions, for example, where erosion exposes relatively favorable subsoils (*13, 52, 62*). The extent to which this occurs globally is difficult to assess given the limited research on soil erosion-productivity.

9. Perhaps most important, the spacial relationships and variability of soils within the landscape have generally been ignored in soil erosion studies. Daniels and associates (*29, 31, 32*) concluded that few if any studies on soil erosion-productivity have adequately accounted for the effect of natural soil variability in erosional landscapes on soil productivity and that the effect of erosion on productivity in the southern Piedmont has been over-estimated, possibly by 50 percent or more. Miller and associates (*74*), using geostatistical techniques, showed that soil properties and wheat growth were influenced by position on the landscape and relationship to their neighbors and concluded that surface soil thickness was not spatially correlated with crop growth. Cassel and Fryrear (*23*) concluded, "Having illustrated the complexities of the erosion-productivity relationships for most landscape situations, it is unlikely that a definitive relationship between soil productivity (as measured by crop yield) and the associated cumulative accelerated soil erosion will be forthcoming for many of our soils, at least at our current level of understanding of landscape processes."

The erosion-productivity relationship from the time accelerated erosion began (Y3 in Figure 1) to the present (Y2 in Figure 1) is difficult to determine and unlikely to be known. Even with careful documentation of spatial variability of soil properties and erosion processes, other problems, not the least of which is identifying and quantifying the soil properties that regulate productivity, would preclude knowing this relationship.

Predicting Future Effects of Soil Erosion on Crop Productivity

Clearly, soils vary in their productivity response to erosion and managed inputs. Therefore, it is conceivable that a number of possible curves exist for the same soil and that a number of soils would respond in similar ways. Possibilities range from complete elimination of soil productivity through the complete loss of soil, to increases in soil productivity because of deposition or the exposure of more favorable subsoil, to no measurable change because of highly buffered soils or the establishment of an equilibrium between losses and gains.

Some efforts have been made to quantify erosion's effects on future productivity through artificial soil removal (*71*) or addition (*73*), by measuring yield relationships where gradients of topsoil thickness occur (*22*), and by analogy. Analogous studies include those in which soil properties and crop yields are measured on a number of sites and, by simplifying assumptions, related to erosion-productivity (*79, 84*). The latter studies suffer from the problems of spatial variation and separation of landscape effects from erosional effects, as well as allowing for technological innovations that might enhance crop productivity.

Determining the effects of erosion on future productivity is obviously a problem of prediction. The choice of prediction techniques depends on the scale of the problem being addressed as well as the level of resource information (data bases) available for the analysis. Existing data bases provide considerable utility in addressing certain issues for which estimates of the erosion-productivity relationship are important.

Development of Predictive Capabilities

Attempts to predict erosion's effects on soil productivity have recognized the limitations discussed above in that the available experimental data fail to provide any mechanistic understanding of the erosion-productivity relationship. Research has established relationships between crop yields and topsoil depth, soil depth, and various linear combinations of soil properties via regression analysis, and to some extent these have been used as proxies or approximations. However, their use as proxies for productivity is certainly limited for reasons discussed earlier. Most importantly, soil properties are spatially variable within a landscape, and the relationships determined in field experiments are often unique to a soil, its location, and the use and management of inputs imposed by the study.

Consider, for example, topsoil depth. Early research that established a relationship between crop yield and topsoil depth formed a basis for prediction of erosion's effect of soil productivity (*115*). Topsoil depth-soil productivity relationships have been described for a number of soils over the last 50 years (*8, 62, 108*), and research relating crop yield to topsoil depth continues (*12, 22, 73*). Therefore, topsoil depth has served as a proxy for soil productivity. Given estimates of soil removal by erosion, estimates of the erosion-productivity relationships are possible, but certainly limited beyond the scope of the experimental conditions under which they were developed.

In the last decade, modeling has taken on a significant role in estimating erosion's impact on soil productivity. A major reason for quantifying the erosion-productivity relationship for soils is to assess the status and trends in soil resources throughout the world. As indicated earlier, this soil resource assessment is legislated in the United States; assessment also is a priority of those concerned with soil degradation elsewhere the world, including the Food and Agriculture Organization of the United Nations and the World Bank.

The assessment of productivity changes because of erosion is a formidable task; it has four basic requirements that must be known or estimated: (1) the inherent productivity of soil, (2) the rate of change in productivity due to erosion, (3) the rate at which erosion occurs, and (4) the areal extent and distribution of soils (spatial variability) (*23, 88*). What constitutes an acceptable prediction method depends upon the problem being addressed and the acceptability of the assumptions underlying the analysis.

The models developed to date to estimate erosion's effect on soil productivity (*27, 35, 53, 89, 94, 103, 114, 120, 125, 127*) have generally adopted existing erosion prediction models, primarily the USLE, to generate soil removal rates for the prediction of productivity changes due to erosion. The basic approach has been to calculate soil productivity (or crop yield) at some point in time (t), alter soil properties in proportion to estimated erosion, recalculate soil productivity of the eroded soil, and repeat this process for some time period or level of soil erosion. Thus, a curve, such as that illustrated in figure 1, could be generated for a given soil under a given set of inputs. The models have differed significantly in their approach to assessing productivity. A key, then, to the success of any model is the degree to which it can approximate productivity and simulate the effects of erosion on the properties that regulate it.

Assessing Soil Productivity

Nix (*78*) described land evaluation in terms of biological or crop productivity as involving the translation of primary data into a quantitative expression of site quality that must include a productivity term. He recognized three approaches: (1) the analogue or transfer-by-analogy approach; (2) the site-factor approach, which seeks to relate key parameters to biological productivity within a given environment where yield is described by a multiple regression equation; and (3) the system-analysis-and-simulation approach, which is concerned with resolution of a complex system into simple component processes that are synthesized into a mathematical model of the whole system.

The analogue methods are based on general land and soil classifications and are dominant. Nix considered site-factor methods to be restricted to certain environments and limited in terms of temporal and spatial extrapolation because they are static representations of a dynamic system. Systems analysis and simulation offer promise to predict the performance of a stated crop genotype at any location, given specified data.

Huddleston (*57*) stated, "Productivity evaluation in the United States has involved a blend of qualitative and quantitative rating models, as well as inductive and deductive thought processes." While evaluation of soil productivity was the primary reason for initiation of soil surveys in the United States, interest shifted in the 1940s and 1950s to the assemblage of crop yield data for soil types in lieu of judging potential productivity based on soil properties in the soil survey.

Ableiter (*1*) described numerical ratings as belonging to two types, inductive ratings derived from inferred effects of various soil and land properties on plant growth and deductive ratings deduced from crop yield records. The inductive ratings are of the parametric type of land evaluation described by Marbutt (*69*). These differ from the site-factor type discussed by Nix (*78*), which are attempts to statistically relate plant growth to measurable properties of soil, physiography, and climate. The most commonly cited example of productivity ratings is the Storie Index (*111, 112, 113*), the first completely inductive attempt to develop quantitative criteria for evaluating soil productivity (*57*). The method is based on soil characteristics that govern the land's potential use and productive capacity (*113*). Nelson and associates (*77*) developed a land productivity index similar to the revised Storie Index (*113*) with the addition of a factor for rainfall. Kiniry and associates (*59*) developed a productivity index (PI) based on the assumption that soil is a major determinate of crop yield because of the environment it provides for root growth, with other factors, including climate, management, and plant genetic potential.

Huddleston (*57*) reviewed the development and use of soil productivity ratings in the United States. Riquier (*98*) summarized parametric methods of soil and land evaluation and discussed a number of methods developed around the world. He listed four possible forms of parametric methods as additive, additive and subtractive, multiplicative, or a more complex equation. Of the many methods that exist, none is preeminent. The advantages and disadvantages of parametric methods cited by Riquier (*98*) are listed in table 2.

Table 2. Advantages and disadvantages of parametric methods of soil and land evaluation (98).

Advantages

1. Each factor and its interactions must be considered.
2. Subjectivity is eliminated since the indices and equations can be standardized.
3. Such a method can be of universal application if factors are well chosen.
4. With the addition of plant parameters, suitability can be determined for each crop separately.
5. A soil potentiality classification can be obtained if plant parameters are ignored.
6. Management factors can be incorporated to arrive at various potentials. The method is very simple and can be repeated successively.
7. The method is quantitative.
8. The method is adapted for computers and data banks to allow the quantitative evaluation of potential in a given area.

Disadvantages

1. Certain factors are difficult to evaluate and are poorly documented.
2. The method is empirical and an approximation. Interactions are not well known.
3. The elaboration of the standard values is experimental and based on comparisons that are themselves subjective.
4. Selection and separation of independent factors is difficult.
5. Missing data must be approximated. This is a problem common to all methods.
6. The mathematical model may hide or hinder a comprehension of the true process. Alternatively it may prove helpful.
7. Comparison between regions is difficult. It is necessary to return to a comparison of limiting factors.

Deductive ratings of productivity based on crop yield have been developed in many states (*7, 11, 16, 43, 44, 51, 70, 82, 101*). They consist of yield tables or empirical productivity indexes derived from crop yields (*57*). While yield-based ratings have the advantage of removing the need for subjective interpretation of soil properties (*1*), lack of sufficient, reliable data is a major limitation (*57*).

The site-factor or soil-site studies, popular in forest soils research, have met with limited success (*21, 25*). Site-factor studies for such crops as corn and soybeans, in which crop yields were regressed on a variety of soil, topographic, and climatic variables, are of limited value to broad areas or regions of soils (*56, 95, 102, 107*). The study by Reich and associates (*95*) illustrates some of the limitations of soil-site studies. Their study included a field survey of 220 soybean (*Glycine max* (L.) Merr.) plots in 12 counties in the Coastal Plain of North Carolina. Some 38 soil, climate, management, and plant variables were measured, with soil variables measured for each horizon. Table 3 gives the independent variables in the reduced model for prediction of yield for each year of the study. The number of predictors is large, and both the number and the independent variables vary from year to year. Results are likely to be specific to the area of study and may include terms of little value elsewhere. While not valuable on a broad scale, some factors recur in these studies as important to crop growth, including soil depth, texture, drainage, stoniness, and fertility.

Hagen and Dyke (*53*) developed a yield/soil loss simulator as part of the 1980 RCA appraisal. Their analysis was the first attempt to use a nationally consistent data base and model to evaluate the effects of soil erosion. Yields, obtained from county soil surveys and normalized to a common base year, 1974, were modeled as a function of soil and other properties using multiple regression analysis. Using data from six different sources, this model was applied to land in the United States for the 1980 RCA appraisal.

Another approach to land evaluation of crop productivity is the use of simulation models described by Tanji (*116*). A variety of models are presently available or under development. While a review of simulation models is beyond the scope of this chapter, some attributes of simulation models and their usefulness to the problem at hand are informative.

Simulation models use a systems analysis approach: a complex system is resolved into simple component processes and synthesized into a mathematical model of whole system (*78*). They vary in complexity from simple models (*110*) to very complex models, such as those developed by Shaffer and associates (*104*). More recent simulation models are linked compilations of existing so-called "state-of-the-art" component models (*125, 127*). These models are large, complex, and have high data requirements. With appropriate components to simulate processes significantly related to soil erosion and crop production, these models

Table 3. Independent variables, their coefficients, and t values from a regression model for prediction of soybean yield (95).

*Independent Variables**	*1974* b value	*1974* t value	*Independent Variable*	*1975* b value	*1975* t value
Crop in field during previous rotation	-423.4	-5.15	Trips	-551.7	
pH	495.3	-3.55	Seed	125.7	
Magnesium	-707.6	-3.55	Planting date	-32.9	
Plant K (%)	293.0	3.01	pH	424.0	
Trips across field before planting	-166.0	-2.85	Plant dry weight (g)	5.4	
Cation exchange capacity-2*	-129.0	-2.76	Magnesium-2*	-1,298.3	
Years since limed	-86.0	-2.60	Slit-3*	18.8	
Aluminum saturation-2*	7.4	2.59	Clay-3*	22.2	
Calcium-2*	213.7	2.46	Intercept	2,600.6	
Rainfall in July and August	-10.3	-2.33	Cation exchange capacity	114.2	
Cation exhange capacity-3*	111.6	2.02	Magnesium	210.3	
Silt-3*	19.5	2.01			
Soil volume weight (g/cc)	-1,202.9	-1.98	**$R^2 = 0.86$**		
Sand-3*	12.8	1.48			
Calcium-3*	-80.0	-1.37			
Intercept	641.8	0.47			
$R^2 = 0.71$					

*Ending number referes to soil horizon.

Table 4. Potential soil productivity losses from erosion in the United States as estimated using the EPIC model (93).

	Productivity Loss in the 100th Year					
	Sheet and Rill Erosion			Wind Erosion		
Region	*Percent*	*Equivalent Acres*	*Gross Product*	*Percent*	*Equivalent Acres*	*Gross Product*
Northeast	7.1	1,108	330	†	3	1
Lake States	0.9	424	124	0.7	255	47
Corn Belt	3.5	3,483	961	†	6	2
Northern Plains	0.6	417	95	0.3	192	35
Appalachia	4.7	883	232	†	†	†
Southeast	1.3	195	52	†	†	†
Delta	1.6	304	72	†	†	†
Southern Plains	0.2	71	16	2.1	573	161
Mountain	0.4	443	15	1.4	442	74
Pacific	2.3	518	74	0.2	77	15
Total	1.8	7,428	1,972	0.5	1,548	335

*Productivity loss is computed from EPIC simulations. Equivalent acre loss is computed as the summation of percent loss of productivity in 100 years times total acres.
†Less than 0.5 percent, .05 million acres, and 0.5 million dollars.

have been used to evaluate soil and land productivity. A simulation model, the Erosion-Productivity Impact Calculator (EPIC) (*126*), specifically designed to estimate the effects of erosion on productivity, was used in the 1985 RCA analysis to project long-term effects of erosion on productivity in the United States. While a strong advantage of simulation models is found in their ability to simulate the many processes affecting crop yield and incorporate present levels of knowledge, knowledge of the processes and necessary data are available for only a limited number of soils (*96*), and calibration of models in different environments is difficult.

The best indicator of soil productivity is crop yield. It is available from farmers' long-term records, crop yield surveys, and plot experiments (*81*). Yield data are not available for all locations, are spatially and temporally variable, and depend on the management and level of technology used. In addition, even a well-documented yield history will not necessarily be indicative of future effects of soil erosion. The above reasons motivated a search for methods to evaluate land and soil productivity.

Long-Term Effects of Erosion on Productivity

Four major studies have addressed the long-term effects of erosion on soil productivity in the United States (*28, 53, 90, 93*). According to each analysis, productivity losses, aggregated over all cropland acres, will not be great over the next 100 years if present erosion rates continue and managed inputs are available. Hagen and Dyke (*53*) concluded from their regression analysis that soil erosion would reduce productivity in the United States 8 percent over the next 100 years. Crosson and Stout (*28*), based on the regression of soil loss on a trend value for yield, concluded that soil erosion reduced the yield trend for corn and soybeans by 1.5 and 2.0 percent, respectively, over the 30-year period from 1950 to 1980. Pierce and associates (*90*) predicted a 4 percent productivity decline due to sheet and rill erosion in the Corn Belt over the next 100 years. Putnam and associates (*93*), using the EPIC model of Williams and associates (*127*) for the second RCA appraisal (*122*), estimated the loss of productivity resulting from wind and water erosion in 100 years for the United States to be 2.3 percent and ranging from 0.7 percent to 7.1 percent, depending on farming region (Table 4). Colacicco and associates (*26*) also used the EPIC model to predict a yield reduction nationally of 4.6 percent for corn over the next 100 years. Their prediction of a 4.2 percent reduction in corn production throughout the Corn Belt is consistent with the other estimates.

Why such low projections for erosion impacts on productivity? Part of the answer lies in the hypothesis that almost two-thirds of U.S. cropland will probably suffer no future yield loss over the next 100 years (*26*). This projection is supported by the results of Pierce and associates (*90*). Although projected productivity losses increase with increasing erosion, a large portion of the Corn Belt does not erode at significant rates (Figure 2). This characteristic, in combination with the occurrence of deep, productive soils with low vulnerabilities to productivity losses from erosion, result in the low estimates of productivity losses in the Corn Belt region of the United States (*90*). However, this does not mean that erosion has little impact on productivity. On the contrary, productivity losses can be high on many soils, particularly where erosion rates are high and soils are shallow or have subsoils unfavorable for root growth (*90*). However, the areal extent of these soils is limited in the Corn Belt and losses were masked in the aggregate analysis. Projected losses in productivity fail to include other effects of erosion on productivity. The analysis also fails to consider the potential pressure placed on the land experiencing no productivity loss once productivity declines on the land.

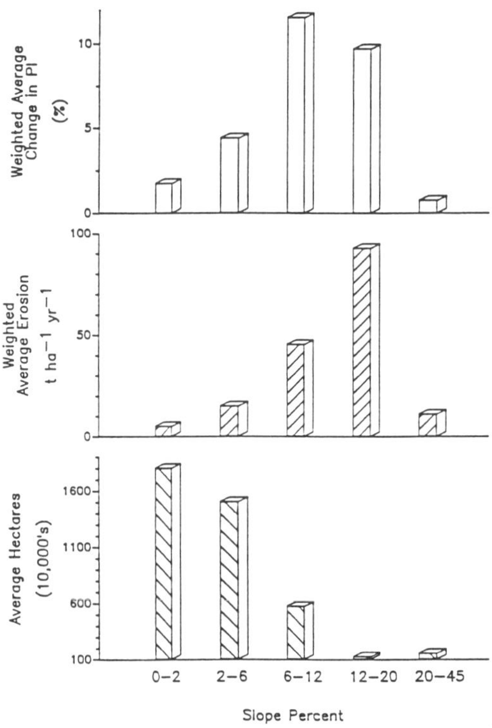

Figure 2. Distribution of weighted average hectares in cropland, erosion rate, and change in productivity (PI) for soils in the Corn Belt region of the United States (*88*).

Uncertainties and Future Considerations

The erosion-productivity relationship for most soils is not known and its measurement is confounded by scales of space and time. Projections about the effects of erosion on soil productivity are difficult, if not impossible, to verify. The existing models are difficult to validate, and the available data bases needed for assessment are grossly inadequate. Erosion estimates are uncertain and not available for many regions of the world. Spatial variation and landscape effects have not been adequately addressed in most erosion-productivity research. It is difficult to separate erosion from other degradation processes. We have yet to set critical limits for physical or economic degradation. Prediction is further complicated by the fact that future conditions, technologies, and resource demands are uncertain.

There is little doubt that considerable research must be initiated to address the unresolved issues regarding soil degradation by erosion. Lal (*62*) and Cassel and Fryrear (*23*) offer important priorities for research on this problem; these will not be repeated here. Instead, I offer three important issues for consideration.

First, agreement must be reached on a credible measuring device. It is not sufficient to continue traditional site-specific, short-term experiments commonly reported in the erosion-productivity literature. While there is a strong need to document the loss in soil productivity from erosion (as well as its potential for recovery), the site-specific and short-term nature of erosion studies limit their utility. It is clear from 50 years of erosion studies that what is not needed is more of these studies. What is needed are (a) models capable of predicting the spatial and temporal distribution of erosion processes and crop productivity,

(b) extensive resource data bases adequate to meet the input needs for these models, and (c) an extensive program to monitor changes in the soil resource base. This would require a clear identification of the soil properties that regulate productivity and a coordinated research and monitoring program designed to support the models. The standardization of field and laboratory methodologies would allow the establishment of a minimum data set for evaluating erosion effects on soil productivity worldwide.

Second, world and national policies that affect and determine land use decisions, resource allocations, and research and program priorities must be identified and monitored. Frankly, public policies that wholly embraced conservation and sustainable development principles would diminish the need to know future trends in soil resources. Simply stated, the policies that affect land use decisions and resource allocations must provide the economic and social incentives necessary for landowners to apply conservation and stewardship principles in their land use decisions.

Third, we must direct considerable efforts toward the restoration and reclamation of degraded land, particularly in the areas where soil resources are limited (developing countries). "Within the usual context of erosion and soil formation," states Wolman (*131*), "relatively little attention is given to recovery or renewal." We must begin to determine the extent to which erosion-induced productivity loss is reversible and under what conditions it would be better to intensify productivity on high quality land than to invest in restoration.

REFERENCES CITED

1. Ableiter, J. K. 1937. *Productivity ratings in the soil survey report.* Soil Science Society of America Proceedings 2: 415-427.
2. Adams, W. E. 1949. *Loss of topsoil reduces crop yields.* Journal of Soil and Water Conservation 4: 130.
3. American Society of Agricultural Engineers. 1985. *Erosion and soil productivity.* St. Joseph, Michigan.
4. Bachtell, M. A., C. J. Willard, and G. S. Taylor. 1956. *Building fertility in exposed soil.* Research Bulletin No. 782. Ohio Agricultural Experiment Station, Wooster.
5. Barnett, A. P., J. R. Carreker, F. Abruna, W. A. Jackson, A. E. Dooley, and J. H. Holladay. 1972. *Soil and treatment losses in runoff with selected cropping treatments on tropical soils.* Agronomy Journal 64(3): 391-395.
6. Batchelder, A. R., and J. N. Jones. 1972. *Soil management factors and growth of Zea Mays L. on topsoil and exposed subsoil.* Agronomy Journal 64: 648-652.
7. Bell, F. F., G. J. Buntley, and P. Denton. 1981. *Yield estimates for the major crops grown on the soils of middle and east Tennessee.* Bulletin No. 604. Tennessee Agricultural Experiment Station, Knoxville.
8. Bennett, H. H. 1939. *Soil conservation.* McGraw-Hill Book Co., New York, New York. 993 pp.
9. Black, A. L., and B. W. Greb. 1968. *Soil reflectance, temperature, and fallow water storage on exposed subsoils of a brown soil.* Soil Science Society of America Proceedings 32: 105-109.
10. Blakie, P. 1985. *The political economy of soil erosion in developing countries.* Longman Scientific and Technical, London, England.
11. Bone, S. W., and L. D. Norton. 1981. *Ohio soils with yield data and productivity index.* Bulletin No. 685. Ohio State University Cooperative Extension Service, Wooster.
12. Bramble-Brodahl, M., M. A. Fosberg, D. J. Walker, and A. L. Falen. 1985. *Changes in soil productivity related to changing topsoil depths on two Idaho Palouse soils.* In R. F. Follett and B. A. Stewart [editors] *Soil Erosion and Crop Productivity.* Soil Science Society of America, Madison, Wisconsin. pp. 18-27.
13. Brink, R. A. 1977. *Soil deterioration and the growing world demand for food.* Science 197: 625-630.
14. Brown, L. R. 1978. *The worldwide loss of cropland. Worldwatch Institute,* Washington, D.C.
15. Brown, L. R. 1984. *The global loss of topsoil.* Journal of Soil and Water Conservation 39: 162-165.
16. Buntley, G. J., and F. F. Bell. 1976. *Yield estimates for major crops grown on soils of west Tennessee.* Bulletin No. 561. Tennessee Agricultural Experiment Station, Knoxville.
17. Buringh, P. 1981. *An assessment of losses and degradation of productive agricultural land in the world.* Working Group on Soils Policy. Food and Agriculture Organization, Rome, Italy.
18. Byalyy, A. M., and T. V. Azovtseva. 1964. *Changes in erosion processes on fields used in a soil protecting crop rotation.* Soviet Soil Science 3: 298.
19. Carlson, C. W., and D. L. Grunes. 1958. *Effect of fertilization on yields and nutrient content of barley.* Soil Science Society of America Proceedings 22: 140-145.
20. Carlson, C. W., D. L. Grunes, J. Alessi, and G. A. Reichman. 1961. *Corn growth as affected by applications of fertilizer and manure.* Soil Science Society of America Proceedings 25: 44-47.
21. Carmean, W. H. 1975. *Forest site quality evaluation in the United States.* Advances in Agronomy 27: 209-269.
22. Carter, D. L., R. D. Berg, and B. J. Sanders. 1988. *The effect of furrow irrigation on crop productivity.* Soil Science Society of America Journal 49: 207-211.
23. Cassel, D. K., and D. W. Fryrear. 1990. *Evaluation of productivity changes due to accelerated soil erosion.* In W. E. Larson, G. K. Foster, R. R. Allmaras, and C. M. Smith [editors] *Research Issues in Soil Erosion/Productivity.* University of Minnesota, St. Paul.
24. Clark, E. H., II, J. A. Haverkamp, and W. Chapman. 1985. *Eroding soils: The off-farm impacts.* The Conservation Foundation, Washington, D.C.
25. Coile, T. S. 1952. *Soil and the growth of forests.* Ad-

vances in Agronomy 4: 329-398.

26. Colacicco, D., T. Osborn, and K. Alt. 1989. *Economic damage from soil erosion.* Journal of Soil and Water Conservation 44: 35-39.
27. Craft, E. M., S. A. Carlson, and R. M. Cruse. 1985. *A model of erosion and subsequent fertilization impacts on soil productivity.* In *Erosion and Soil Productivity.* American Society of Agricultural Engineers, St. Joseph, Michigan. pp. 143-151.
28. Crosson, P. R., and A. J. Stout. 1983. *Productivity effects of cropland erosion in the United States.* Resources for the Future, Washington, D.C.
29. Daniels, R. B., and G. D. Bubenzer. 1990. *Landscape processes and soil erosion and soil productivity.* In W. E. Larson, G. K. Foster, R. R. Allmaras, and C. M. Smith [editors] *Research Issues in Soil Erosion/Productivity.* University of Minnesota, St. Paul.
30. Daniels, R. B., J. W. Gilliam, D. K. Casses, and L. A. Nelson. 1985. *Soil erosion and landscape position in the North Carolina Piedmont.* Soil Science Society of America Journal 49: 991-995.
31. Daniels, R. B., J. W. Gilliam, D. K. Casses, and L. A. Nelson. *Quantifying the effects of past soil erosion on present productivity.* Journal of Soil and Water Conservation 42: 183-187.
32. Daniels, R. B., J. W. Gilliam, D. K. Casses, and L. A. Nelson. 1989. *Soil erosion has limited effect on field scale crop productivity in the Southern Piedmont.* Soil Science Society of America Journal 53: 917-920.
33. Dregne, H. 1982. *Historical perspectives of accelerated erosion and effect on world civilizations.* In B. L. Schmidt [editor] *Determinants of Soil Loss Tolerance.* Special Publication 45. American Society of Agronomy, Madison, Wisconsin. pp. 1-14.
34. Dudal, R. 1982. *Land degradation in a world perspective.* Journal of Soil and Water Conservation 37: 345-349.
35. Dyke, P. T., and E. O. Heady. 1985. *Assessment of soil erosion and crop productivity with economic models.* In R. F. Follett and B. A. Stewart [editors] *Soil Erosion and Crop Productivity.* Soil Science Society of America, Madison, Wisconsin. pp. 105-117.
36. Eck, H. V. 1969. *Restoring productivity on Pullman silty clay loam subsoil under limited moisture.* Soil Science Society of America Proceedings 33: 578-581.
37. Eck, H. V., and R. H. Ford. 1962. *Restoring productivity on exposed subsoils.* Journal of Soil and Water Conservation 17: 274-275.
38. Eck, H. V., V. L. Hauser, and R. H. Ford. 1965. *Fertilizer needs for restoring productivity on Pullman silty clay loam after variable degrees of soil removal.* Soil Science Society of America Proceedings 29: 209-213.
39. Eckholm, E. P. 1976. *Losing ground: Environmental stress and world food prospects.* Norton, New York.
40. El-Swaify, S. A., E. W. Dangler, and C. L. Armstrong. 1982. *Soil erosion by water in the tropics.* Research Extension Series, HITAHR. College of Tropical Agriculture and Human Resources, University of Hawaii, Honolulu.
41. El-Swaify, S. A., W. C. Moldenhauer, and A. Lo. 1985. *Soil erosion and conservation.* Soil Conservation Society of America, Ankeny, Iowa.
42. Englestead, O. P., W. D. Shrader, and L. C. Dumenil. 1961. *The effect of surface soil thickness on corn yields: I. As determined by a series of field experiments on farmer-operated fields.* Soil Science Society of America Proceedings 25: 494-497.
43. Fehrenbacker, J. B., R. A. Pope, I. J. Jansen, J. D. Alexander, and B. W. Ray. 1978. *Soil productivity in Illinois.* Circular No. 1156. University of Illinois Cooperative Extension Service, Urbana.
44. Fenton, T. E., E. R. Duncan, W. D. Shrader, and L. C. Dumenil. 1971. *Productivity levels of some Iowa soils.* Special Report 66. Iowa Agriculture and Home Economics Experiment Station and Cooperative Extension Service, Ames.
45. Follett, R. F., and B. A. Stewart, editors. 1985. *Soil erosion and crop productivity.* Soil Science Society of America, Madison, Wisconsin.
46. Food and Agriculture Organization. 1977. *Assessing soil degradation.* Soils Bulletin 34:83. Rome, Italy.
47. Food and Agriculture Organization. 1978. *Methodology for assessing soil degradation.* Report on Food and Agriculture Organization Expert Consultation. Rome, Italy.
48. Food and Agriculture Organization. 1979. *A provisional methodology for assessing soil degradation.* Rome, Italy.
49. Frye, W. W., S. A. Ebelhar, L. W. Murdock, and R. L. Blevins. 1982. *Soil erosion effects on properties and productivity of two Kentucky soils.* Soil Science Society of America Proceedings 46: 1,051-1,055.
50. Grandt, A. F. *Reclaiming mined land in Illinois for row crop production.* Journal of Soil and Water Conservation 33: 242-243.
51. Gray, F., E. C. Nance, and J. H. Stiegler. 1979. *Productivity of benchmark and key soils in Oklahoma.* Bulletin No. B-740. Oklahoma Agricultural Experiment Station, Stillwater.
52. Grosse, B. 1967. *The productivity of severely eroded Parabraunerde formed from loess under moderate climatic conditions.* Transactions, Eighth International Congress Soil Science 162(2): 729.
53. Hagen, L. L., and P. T. Dyke. 1980. *Merging data from disparate sources.* Agricultural Economics Research 32(4): 45-49.
54. Hays, O. E., C. E. Bay, and H. H. Hull. 1948. *Increasing production on an eroded loess-derived soil.* Journal of American Society of Agronomy 40: 1,061-1,069.
55. Heilman, M. D., and J. R. Thomas. 1961. *Land leveling can adversely affect soil fertility.* Journal of Soil and Water Conservation 16: 71-72.
56. Henao, J. 1976. *Soil variables for regressing Iowa corn yields on soil, management, and climatic variables.* Ph.D. thesis. Iowa State University, Ames. 315 pp.
57. Huddleston, J. H. 1984. *Development and use of soil productivity ratings in the United States.* Geoderma 32: 297-317.
58. Kellogg, C. E. 1975. *Agricultural development: Soil, food, people, work.* Soil Science Society of America, Madison, Wisconsin.
59. Kiniry, L. N., C. L. Scrivner, and M. E. Keener. 1983. *A soil productivity index based upon predicted water depletion and root growth.* Research Bulletin 1051. Missouri Agricultural Experiment Station, Columbia.
60. Krantz, B. A. 1958. *Soil survey interpretation-interpretation of soil characteristics in soil management.* Soil Science Society of America Proceedings 23: 155-156.
61. Lal, R. 1980. *Soil erosion as a constraint to crop production.* In *Priorities for Alleviating Soil-Related Constraints to Food Production in the Tropics.* The International Rice Research Institute and New York College of Agriculture and Life Sciences, Cornell University, Ithaca, New York. 468 pp.
62. Lal, R. 1987. *Effects of soil erosion on crop productivity.* CRC Critical Reviews in Plant Sciences 5(4): 303-367.

63. Lal, R. 1990. *Soil erosion and land degradation: The global risks.* In R. Lal and B. A. Stewart [editors] *Soil Degration.* Advances in Agronomy 11: 129-172.
64. Langdale, G. W., and W. D. Shrader. 1982. *Soil erosion effects on soil productivity of cultivated cropland.* In B. L. Schmidt, R. R. Allmaras, J. V. Mannering, and R. I. Papendick [editors] *Determinants of Soil Loss Tolerance.* Special Publication No. 45. American Society of Agronomy, Madison, Wisconsin. pp. 41-51.
65. Larson, W. E., F. J. Pierce, and R. H. Dowdy. 1983. *The threat of soil erosion to long-term crop production.* Science 219: 458-465.
66. Latham, E. E. 1940. *Relative productivity of the A horizon on the Cecil sandy loam and the B and C horizons exposed by erosion.* Journal of American Society of Agronomy 32: 950-954.
67. Lowdermilk, W. C. 1975. *Conquest of the land through 7,000 years.* Information Bulletin 99. U.S. Department of Agriculture, Washington, D.C.
68. Lyles, L. 1975. *Possible effects of wind erosion on soil productivity.* Journal of Soil and Water Conservation 30:279-283.
69. Mabbutt, J. A. 1968. *Review of concepts of land classification.* In G. A. Stewart [editor] *Land Evaluation.* The Macmillan Company of Australia, Melbourne. pp. 11-28.
70. Malo, D. D., and F. C. Westin. 1978. *Rating South Dakota soils according to productivity.* Bulletin No. 657. South Dakota Agricultural Experiment Station, Brookings.
71. Mbagwu, J. S., R. Lal, and T. W. Scott. 1984. *Effects of desurfacing of Alfisols and Ultisols in Southern Nigeria.* I. Crop performance. Soil Science Society of America Journal 48: 828.
72. Meyer, L. D., A. Bauer, and R. D. Heil. 1985. *Experimental approaches for quantifying the effects of soil erosion on productivity.* In R. F. Follett and B. A. Stewart [editors] *Soil Erosion and Crop Productivity.* Soil Science Society of America, Madison, Wisconsin.
73. Mielke, L. N., and J. S. Sheppers. 1986. *Plant response to topsoil thickness and to fertilizer on eroded loess soil.* Journal of Soil and Water Conservation 41: 59-63.
74. Miller, M. P., M. J. Singer, and D. R. Nielsen. 1988. *Spatial variability of wheat yield and soil properties on complex hills.* Soil Science Society of America Journal 51: 1,563-1,565.
75. National Association of Conservation Districts. 1980. *Soil degradation: Effects on agricultural productivity.* Interim Report No. 4. Washington, D.C.
76. Neal, O. R. 1943. *The influence of soil erosion on fertility losses and on potato yield.* American Potato Journal 20: 57-64.
77. Nelson, et al. 1963. *Detailed land classification-island of Ohau.* Land Study Bureau Bulletin 3. University of Hawaii, Honolulu. 141 pp.
78. Nix, H. A. 1968. *The assessment of biological productivity.* In G. A. Stewart [editor] *Land Evaluation.* The Macmillan Company of Australia, Melbourne. pp. 77-87.
79. Nizeyimana, E., and J. R. Olson. 1988. *Chemical, mineralogical, and physical property differences between moderately and severely eroded Illinois soils.* Soil Science Society of America Journal 52: 1,740-1,748.
80. Odell, R. T. 1950. *Measurement of productivity of soils under various environmental conditions.* Agronomy Journal 42: 282-292.
81. Odell, R. T. 1958. *Soil survey interpretation: Yield prediction.* Soil Science Society of America Proceedings 22: 157-160.
82. Odell, R. T., and W. R. Oschwald. 1970. *Productivity of Illinois soils.* Circular 1016. University of Illinois Cooperative Extension Service, Urbana.
83. Olson, G. W. 1981. *Soils and the environment: A guide to soil surveys and their applications.* Chapman and Hall, New York, New York. 178 pp.
84. Olson, J. R., and E. Nizeyimana. 1988. *Effects of soil erosion on corn yields of seven Illinois soils.* Journal of Production Agriculture 1: 13-19.
85. Olson, T. C. 1977. *Restoring the productivity of a glacial till soil after topsoil removal.* Journal of Soil and Water Conservation 32: 130-132.
86. Papendick, R. I., D. L. Young, D. K. McCool, and H. A. Krauss. 1985. *Regional effects of soil erosion on crop productivity—The Palouse area of the Pacific Northwest.* In R. F. Follett and B. A. Stewart [editors] *Soil Erosion and Crop Productivity.* Soil Science Society of America, Madison, Wisconsin. pp. 306-338.
87. Phillips, J. A., and E. J. Kamprath. 1973. *Soil fertility problems associated with land forming in the Coastal plain.* Journal of Soil and Water Conservation 28: 69-73.
88. Pierce, F. J. 1987. *Modeling soil productivity changes due to erosion.* Transactions, XIII Congress of International Society of Soil Science: 258-265.
89. Pierce, F. J., W. E. Larson, R. H. Dowdy, and W.A.P. Graham. 1983. *Productivity of soils: Assessing long-term changes due to erosion.* Journal of Soil and Water Conservation 38: 39-44.
90. Pierce, F. J., W. E. Larson, R. H. Dowdy, and W.A.P. Graham. 1984. *Soil productivity in the Corn Belt: An assessment of erosion's long-term effects.* Journal of Soil and Water Conservation 39: 131-136.
91. Pimentel, D., E. C. Terhune, R. Dyson-Hudson, S. Rocherau, R. Sames, E. A. Smith, D. Denman, D. Reifschneider, and M. Shepard. 1976. *Land degradation: Effects on food and energy resources.* Science 194: 149-155.
92. Power, J. F., F. M. Sandoval, R. E. Ries, and S. D. Merrill. 1981. *Effects of topsoil and subsoil thickness on soil water content and crop production on a disturbed soil.* Soil Science Society of America Proceedings 45: 124-129.
93. Putman, J., J. Williams, and D. Sawyer. 1988. *Using the erosion-productivity impact calculator (EPIC) model to estimate the impact of soil erosion for the 1985 RCA appraisal.* Journal of Soil and Water Conservation 43: 321-326.
94. Putman, J., P. Dyke, and G. Wistrand. 1985. *Erosion-productivity index simulator (EPIS).* Economical Rsearch Service, U.S. Department of Agriculture, Washington, D.C.
95. Reich, R. C., E. J. Kamprath, and L. A. Nelson. 1981. *Soil management factors correlated with soybean yields in Southeastern U.S. Coastal plain.* Agronomy Journal 73: 90-95.
96. Reijsberman, F. R., and M. Gordon Wolman. 1984. *Quantification of the effect of erosion on soil productivity in an international context.* International Federation of Institutes for Advanced Study, Analyzing Biospheric Change. Solna, Sweden.
97. Reuss, J. O., and R. E. Campbell. 1961. *Restoring productivity to leveled land.* Soil Science Society of America Proceedings 25: 302-304.

98. Riquier, J. 1974. *A survey of parametric methods of soil and land evaluation.* In *Approaches to Land Classification.* Soils Bulletin 22. Food and Agriculture Organization, Rome, Italy. pp. 47-53.
99. Riquier, J. 1977. *Philosophy of the world assessment of soil degradation and items for discussion.* In *Food and Agriculture Organization Assessing Soil Degradation.* Soils Bulletin. Food and Agriculture Organization, Rome, Italy. pp. 36-38.
100. Rost, C. O. 1940. *The relative productivity of humid subsoils.* Soil Science Society of America Proceedings 4: 281-287.
101. Rust, R. H., and L. D. Hanson. 1975. *Crop equivalent rating guide for soils of Minnesota.* Miscellaneous Report 132. Minnesota Agricultural Experiment Station, St. Paul.
102. Sakar, P. K., O. W. Bidwell, and L. F. Marcus. 1966. *Selection of characteristics for numerical classification of soils.* Soil Science Society of America Proceedings 30: 269-272.
103. Shaffer, M. J. 1985. *Simulation model for soil erosion-productivity relationships.* Journal of Environmental Quality 14: 144-150.
104. Shaffer, M. J., S. C. Gupta, D. R. Linden, J..A.E. Molina, C. E. Clapp, and W. E. Larson. 1983. *Simulation of nitrogen, tillage, and residue management effects on soil fertility.* In *Proceedings, International Conference on State-of-the-Art in Ecological Modelling.* Colorado State University, Fort Collins.
105. Shuman, G. E., and J. F. Power. 1981. *Topsoil management on mined lands.* Journal of Soil and Water Conservation 36: 77-78.
106. Smith, R. M., R. C. Henderson, E. D. Cook, J. E. Adams, and D. O. Thompson. 1967. *Renewal of desurfaced Austin clay.* Soil Science 103: 126-130.
107. Sopher, C. D., and R. J. McCracken. 1973. *Relationships between soil properties, management, practices, and corn yields on South Atlantic Coastal plain soils.* Agronomy Journal 65: 595-599.
108. Stallings, J. H. 1950. *Erosion of topsoil reduces productivity.* Technical Publication No. 98. Soil Conservation Service, U.S. Department of Agriculture, Washington, D.C.
109. Stallings, J. H. 1957. *Soil conservation.* Prentice-Hall, Englewood Cliffs, New Jersey.
110. Stapper, M., and G. F. Arkin. 1980. *CORNF: A dynamic growth and development model for maize* (Zea mays *L.).* Texas Agricultural Experiment Station, Temple. 91 pp.
111. Storie, R. E. 1933. *An index for rating the agricultural value of soils.* Bulletin 556. California Agricultural Experiment Station, Berkeley.
112. Storie, R. E. 1937. *An index for rating the agricultural value of soils.* Bulletin 556 (revised). California Agricultural Experiment Station, Berkeley.
113. Storie, R. E. 1978. *Storie index rating.* Special Publication 3203. Division of Agricultural Science, University of California, Berkeley.
114. Swan, J. B., M. J. Shaffer, W. H. Paulson, and A. E. Peterson. 1987. *Simulating the effects of soil depth and climatic factors on corn yield.* Soil Science Society of America Journal 51: 1,025-1,032.
115. Swanson, E. R., and C. E. Harshbarger. 1964. *An economic analysis of effects of soil loss on crop yields.* Journal of Soil and Water Conservation 19: 183-186.
116. Tanji, K. K. 1981. *Approaches and philosophy of modeling.* In I. K. Iskandar [editor] *Modeling Wastewater Renovation.* John Wiley, New York, New York. pp. 20-41.
117. Thomas, H. L., and D. K. Cassel. 1979. *Land-forming Atlantic Coastal plain soils: Crop yield relationships to soil physical and chemical properties.* Journal of Soil and Water Conservation 34: 20-24.
118. Thomas, J. R., M. D. Heilman, and L. Lyles. 1974. *Predicting nitrogen fertilizer requirements after land leveling.* Agronomy Journal 66: 371-374.
119. Thomas, P. J., T. W. Simpson, and J. C. Baker. 1989. *Erosion effects on productivity of Cecil soils in the Virginia Piedmont.* Soil Science Society of America Journal 53: 928-983.
120. Timlin, D. J., R. B. Bryant, V. A. Snyder, and R. J. Wagenet. 1986. *Modeling corn grain yield in relation to soil erosion using a water budget approach.* Soil Science Society of America Journal 50: 718-723.
121. Uhland, R. E. 1949. *Crop yields lowered by erosion.* SCS-TP-75. Soil Conservation Service, U.S. Department of Agriculture, Washington, D.C.
122. U.S. Department of Agriculture. 1989. *The second RCA appraisal.* Soil, Water and Related Resources on Nonfederal Land in the United States, Analysis of Conditions and Trends. Washington, D.C.
123. U. S. Department of Agriculture and Office of Technology Assessment, U.S. Congress. 1982. *The impact of technology on the productivity of the land.* Washington, D.C.
124. Whitney, R. S., R. Gardner, and D. W. Robertson. 1950. *The effectiveness of manure and commercial fertilizer in restoring the productivity of subsoils exposed by leveling.* Agronomy Journal 42: 239-245.
125. Williams, J. R., C. A. Jones, and P. T. Dyke. 1984. *A modeling approach to determining the relationship between erosion and soil productivity.* Transactions, American Society of Agricultural Engineers 27: 129-144.
126. Williams, J. R., and K. G. Renard. 1985. *Assessment of soil erosion and crop productivity with process models (EPIC).* In R. F. Follett and B. A. Stewart [editors] *Soil Erosion and Crop Productivity.* Soil Science Society of America, Madison, Wisconsin. pp. 67-103.
127. Williams, J. R., K. G. Renard, and P. T. Dyke. 1983. *EPIC: A new method for assessing erosion's effect on soil productivity.* Journal of Soil and Water Conservation 38: 381-383.
128. Williams, J. R., R. R. Allmaras, K. G. Renard, Leon Lyles, W. C. Moldenhauer, G. W. Langdale, L. D. Meyer, and W. J. Rawls. 1981. *Soil erosion effects on soil productivity: A research perspective.* Journal of Soil and Water Conservation 36: 82-90.
129. Winters, E., and R. W. Simonson. 1950. *The subsoil.* Advances in Agronomy 3: 1-92.
130. Wittwer, S. H. 1986. *Crop productivity—research imperatives: A decade of change.* In Martin Gibbs and Carla Carlson [editors] *Crop Productivity—Research Imperatives Revisited.* National Academy Press, Washington, D. C.
131. Wolman, M. G. 1985. *Soil erosion and crop productivity: A worldwide perspective.* In R. F. Follett and B. A. Stewart [editors] *Soil Erosion and Crop Productivity.* Soil Science Society of America, Madison, Wisconsin. pp. 9-21.

7

Adoption of Conservation Tillage and Associated Planting Systems

R. R. Allmaras, G. W. Langdale, P. W. Unger, R. H. Dowdy, and D. M. Van Doren

Conservation tillage was recently defined as a system in which either crop residues are retained on or near the soil surface, a rough soil surface is maintained, or both to control soil erosion and to achieve good soil-water relations (*3, 72*). This is the most widely used abstract definition. Mannering and Fenster (*72*) cited numerous conservation tillage systems, each of which reduced soil erosion compared to a conventional system, which at the time did not have the surface roughness or residue cover necessary to control erosion. The good soil-water relations objective is less quantified than the soil-erosion control objective.

For operational and assessment purposes, forms of conservation tillage-planting are now defined (*23, 43*) according to the amount of residue cover (greater than 30 percent of surface covered) on the soil surface after planting and the proportion of area (a full or partial width of tillage) cultivated. Schertz (*95*) sketched the origin and development of the operational definition based upon surface residue cover. When 30 percent of surface is covered by crop residue, wind or water erosion is expected to decline to 50 percent of that on a bare soil. This operational definition does not include some forms of tillage considered to be conservation tillage in the abstract definition. Such an operational deficiency should not exclude the use of a tillage system that provides soil and water conservation, especially if it proves to be economical and sustaining.

The good soil-water relations objective in the definition of conservation tillage relates to tillage management of both crop residue and soil surface roughness to improve infiltration and water redistribution in soil and to reduce evaporation. Unger and associates (*109*) indicated that the first association of surface residue with water conservation involved erosion control through reduced runoff and that the association between surface residues, enhanced water infiltrated, and reduced evaporation was developed in the late 1930s. Research since 1930 has provided much information about the forms of residue and tillage management and the processes involved in soil-water conservation (*109, 113*).

Soil surface roughness (surface microrelief, random microrelief, random roughness) is not included in the operational definition (*23*). A mere comparison of surface residue cover and random roughness after moldboard plowing, chisel plowing, and no-tillage, all as primary tillages, illustrates that the two properties are not correlated. Larson (*65*) specified a need for random roughness of the interrow to improve infiltration; later work of Burwell and Larson (*15*) demonstrated the impact of random roughness on the "time to runoff" during simulated rain. These researchers also demonstrated that tillage control of random roughness could prevent runoff until as much as 70 percent of the average annual erosion index has accumulated in the north central United States. However, extensive secondary tillage significantly reduced this potential for runoff control. Laflen and associates (*60*) developed algorithms for computing C values (in the USLE) related to random roughness and crop residue, both buried and on the surface. Voorhees and associates (*114*) and Zobeck and Onstad (*127*) provide approximate values of random roughness associated with various tillage implements and their combined use.

Among the first conservation tillage experiments were those conducted jointly in the United States and Canada, starting in 1941, that used stubble mulch to reduce wind erosion and enhance soil-water storage in the semiarid dryland (*77*). An account of machinery developments in stubble mulch farming is provided by Allen and Fenster (*1*).

Conservation tillage evolution in the southeastern United States first succeded in the 1930s in the Georgia Piedmont. A double-cropping system was used to manage crop residues for erosion control (*70*) with a 10-centimeter-wide bull tongue (a rigid shank with a chisel point rather than an attached shovel) on a 17 percent slope. However, crop residue management systems (*67*) were difficult to develop because of technological needs, such as machinery, fertilization, and herbicides (*64*).

Other examples of pioneering research in conservation tillage involve the use of soybean (*Glycine max* L.), wheat (*Triticum aestivum* L.), and corn (*Zea mays* L.) residue to reduce water erosion in Illinois (*111*); development of plow-plant systems in New York (*83*); and introduction of the no-till system for row crops in Virginia (*79*). Field application of stubble mulch took place rapidly in the 1940s, but other forms of conservation tillage in the northern latitudes of the United States did not become operational until the 1970s.

From 1950 to 1970, much research showed significant reductions of wind and water erosion attainable with conservation tillage systems (*124, 126*). By 1982, conservation tillage was used on 25 percent of the tilled land in the United States (*85*). Many factors facilitated and mandated the development of conservation tillage systems. Tractor and machinery design increased the potential for excessive tillage, but also facilitated field operations in the presence of crop residues on the soil surface. Fuel costs became a significant production cost, public alarm heightened regarding soil erosion, and herbicidal weed control reduced the need for frequent tillage. During the early phases of conservation tillage development, serious yield reductions occurred. Since 1970, however, marked improvements have been made in the yield performance in conservation tillage systems. In many cases yields are now superior for conservation tillage systems (*3, 4, 63*). These recent yield trends are indicative of a rapidly advancing technology.

Conservation tillage success in controlling erosion has prompted its use as a primary practice for conservation compliance in the 1985 farm bill. Conservation tillage was already considered the "leading cost effective practice" for soil erosion control within the Soil and Water Resources Conservation Act of 1977 (*110*). Such reliance upon conservation tillage explains the value of the annual survey of adoption conducted since 1982 by the Conservation Technology Information Center.

It is our objective to evaluate these statistics as they relate to research findings and projections in various tillage management regions, originally defined by Allmaras and associates (*3*). Such an evaluation will suggest approaches to be taken in the survey to better characterize the development and use of tillage systems.

Soil conservation problems, including those involving tillage, were detailed on a geographical basis in 1965 (*99*). Based on common climate, topography, soil, land use, and cultural practice for a geographic area, the United States was subdivided into 182 major land resource areas. This geographical classification projected problems and benefits from soil conservation as related to a set of common factors. The concept of a tillage management re-

Table 1. Definitions of conservation tillage used in a national survey of adoption (23).

*Tillage-Planting System**	*Treatment During Harvest to Planting Period*	*Width of Disturbance Over Row*	*Weed Control*	*Abbreviation*
No-till (slot plant)	None†	2-7 cm	Herbicide	NT
Ridge till	None	1/3 of area‡	Combination§	RiT
Strip till	None	1/3 of area#	Combination	ST
Mulch till	Full width‖	Whole area	Combination	MT
Reduced till	Full width‖	Whole area	Combination	ReT

*Conservation tillage is a tillage and planting system that retains at least a 30% cover of crop residue on the soil surface after planting. Residue cover may be a forage crop, winter cover crop, small grain, or row crop. Residue cover can be computed from the weight of residue on the surface (*44*). Reduced till was included in the original survey as an undefined form of conservation tillage that met the 30% cover requirement.

†No disturbance with tillage.

‡Ridges 10 to 15 cm higher than middle of interrow. Tillage done with sweeps or row cleaners; cultivation used to rebuild ridges.

§Herbicide and cultivation combined.

#Tillage in row with rototiller, chisel in the row, or row cleaner.

‖Tillage with chisel, field cultivator, disc, sweep, or special blades. In mulch till, the full-width, primary tillage is performed before planting; in reduced till the full-width, primary tillage is performed simultaneously with planting.

gion (*3, 4*) is an extension and revision of the major land resource areas classification. Tillage management regions attempt to delineate geographic areas in which specific forms of conservation tillage are performed. In both major land resource areas and tillage management regions, aggregation over area must address variation that exists among land resource units. A land resource unit was defined as a "geographic area of land, usually about a thousand hectares in extent, that is characterized by particular patterns and combinations of soil (including slope and erosion), climate, water resources, land use, tillage, and type of farming" (*99*).

To facilitate the evaluations of conservation tillage systems adoption, (a) 11 tillage management regions were delineated (in the United States), within each of which farmer adoption of tillage-planting systems was evaluated; (b) some of the most important factors involved in conservation tillage-planting systems were discussed; and (c) methods to systematize the adoption trends were described and explained.

Conservation Tillage-Planting System

The current operational definition of a conservation tillage-planting system (Table 1) has been used since 1982 when the Conservation Technology Information Center began a national inventory of conservation tillage adoption. The first three forms of conservation tillage are considered to be partial-width systems because, whatever type of tillage is performed, it is not the same over the whole width of the implement (the implement usually spans more than one crop row). One form of the full-width systems (mulch till) ordinarily applies a uniform, overall tillage prior to planting, whereas the other full-width system (reduced till) involves a one-time operation at planting (i.e., no tillage since harvest). Typical seedbed arrangements of surface residue placement and soil disturbance are shown in figure 1. These characterize the impacts of partial and full-width tillage on the crop residue distribution.

Reduced till was defined in the survey as "any other tillage and planting system (other than the four listed first in Table 1) that meets the 30 percent cover rule." Undoubtedly, the term "reduced till" has been used often to include other forms of conservation tillage because the term "reduced tillage" is a vestige of definitions used early in conservation tillage literature. In fact, reduced tillage was recently defined to include the whole range of tillage practices that reduce the number of tillage operations relative to conventional tillage (*102, 120*). Wiese (*120*) also used reduced, limited, and minimum tillage synonomously. Many glossaries use reduced till to describe a one-time, full-width conservation tillage with all tillage delayed until planting time. Reduced tillage is also defined as the use of primary tillage in conjunction with a planting operation to reduce or eliminate secondary tillage

(*5, 101*). Primary tillage is suggested to always be broadcast or full width (*101, 102*). To be used operationally, reduced tillage must leave 30 percent cover after planting.

As mentioned earlier, the operational definition in table 1 makes no mention of surface roughness; the amount of random surface roughness and its value for soil erosion control cannot be implied by the observation of more or less than 30 percent cover of crop residue after planting. For fine-textured soils, random roughness can be estimated from a knowledge of tillage operations/pictorial guides (*127*). In coarse-textured soils, random roughness is often either inconspicuous or ephemeral as related to rainfall energy.

Tillage Management Regions

A tillage management region should be delineated so that technological problems and the potential for development of a conservation tillage system applies within a tillage management region; between tillage management regions, the problems and potential for a conservation tillage system should differ

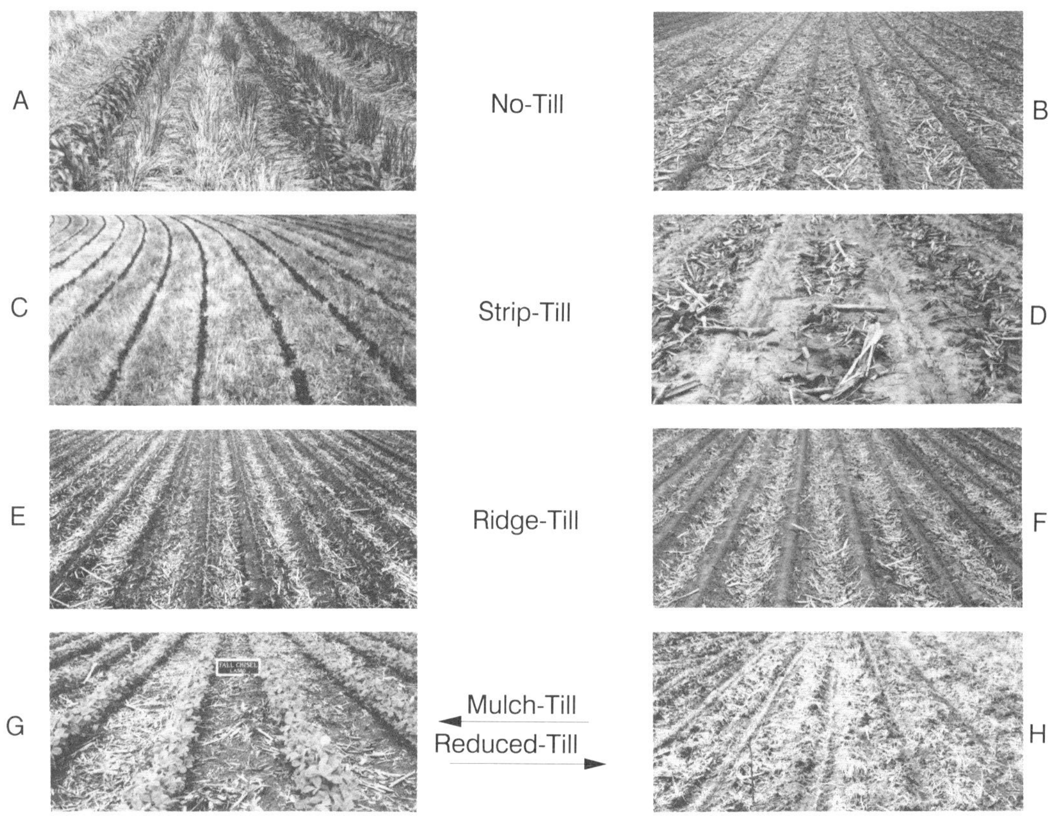

Key	*Crop Sequence*	*State*	*Courtesy*
A	Wheat-Sorghum	Texas	P. W. Unger, USDA/ARS
B	Corn-Soybeans	Nebraska	E. C. Dickey, Univ. of Nebr.
C	Wheat-Soybeans (DC)	Georgia	G. L. Langdale, USDA/ARS
D	Corn-Corn	Nebraska	E. C. Dickey, Univ. of Nebr.
E	Corn-Corn	Minnesota	G. W. Randall, Univ. of Minn.
F	Corn-Soybeans	Minnesota	G. W. Randall, Univ. of Minn.
G	Corn-Soybeans	Minnesota	G. W. Randall, Univ. of Minn.
H	Small Grain-Corn	Nebraska	E. C. Dickey, Univ. of Nebr.

Figure 1. Surface residue and tillage-width characteristcs of the five conservation tillage-planting systems of table 1.

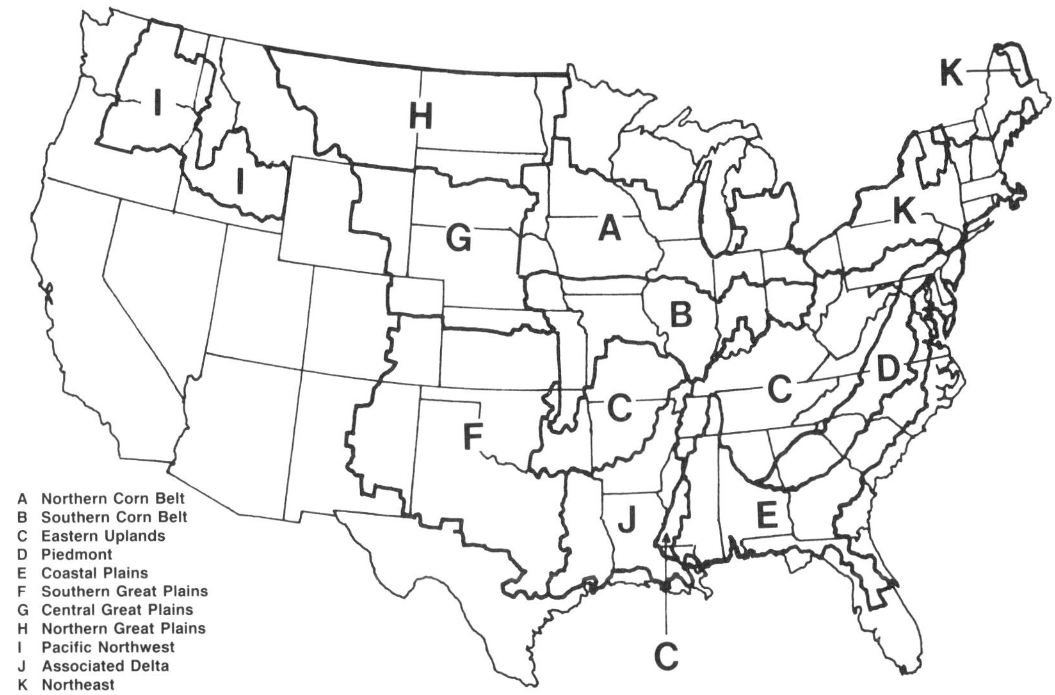

Figure 2. Eleven tillage management regions (TMR) in the conterminous United States.

significantly. Eleven tentative tillage management regions are delineated in figure 2 to systematize technological problems, advancement potential, and benefits. This delineation is revised somewhat compared with those shown earlier (*3, 4*).

Climate (Table 2) and adapted major crops (Tables 3, 4, 5, 6, 7) are two prominent factors that delineate the tillage management regions. The four western-most tillage management regions are dry subhumid to semiarid, as shown by a mean annual precipitation that is less than mean annual pan evaporation. The seven eastern-most tillage management regions are wet subhumid to humid, with small superhumid zones in the Eastern Uplands. These tillage management regions show a mean annual potential evaporation that is less than their mean annual precipitation. In the Great Plains and Corn Belt tillage management regions, there is also a significant gradation of decreasing precipitation effectiveness from east to west. Mean monthly temperatures in January and July, as well as the number of continuous frost-free days, show a gradation of thermal climate among the 10 eastern-most tillage management regions.

Dot charts showing geographic distribution of crops harvested in 1978 (*13*) and the county-by-county summary of conservation tillage-planting systems for each crop (*23*) both show the same adopted major crops for a tillage management region. The percentage of the total hectares (in the conterminous United States) of a crop type planted in each tillage management region (Table 3) demonstrates that corn is concentrated in the Northern/Southern Corn Belt and the Central Great Plains, with a small amount in all other tillage management regions. Soybean is concentrated in the Northern/Southern Corn Belt and Associated Delta, with moderate amounts in other tillage management regions, except the Pacific Northwest, Northeast, and Northern Great Plains, where production is sparse. Small grain (Table 3) includes species with varying areal adaptations. Wheat is concentrated in the four western-most tillage management regions, with a lesser concentration in the eastern part of the Southern/Northern Corn Belt, the Eastern Uplands, and Piedmont. Barley (*Hordeum vulgare* L.) production is concentrated in the Pacific Northwest. Northern Great Plains, and the western portion of the Northern Corn Belt, with some grown in the Piedmont, Southern Great Plains, and Central Great Plains tillage management regions. Rye (*Secale cereale* L.) is grown also in the Coastal Plains, Central Great Plains, Piedmont, and Southern Great Plains tillage management regions. Oats (*Avena sativa* L.) is concentrated in the Northern Great Plains, Central Great Plains, Northern Corn Belt, and Northeast tillage management regions. Although the information is aggregated by states, Daberkow and Gill (*27*) describe the distribution of barley, oats, and the various wheats.

Grain sorghum (*Sorghum bicolor* L.) is concentrated in the Central/Southern Great Plains, with some adaptation in the western parts of the Southern Corn Belt and in the Piedmont tillage management regions. Cotton (*Gossypium hirsutum* L.) is found almost exclusively in the Associated Delta and Southern Great Plains, with minor production in the Coastal Plains tillage management regions.

Among "other crops" (Table 3), there are exclusive areas of production in the tillage management regions as obtained from dot charts of the 1978 Census of Agriculture (*13*). Sunflower (*Helianthus annuus* L.) is almost exclusive to the Northern Great Plains, with some production in adjacent parts of the Northern Corn Belt, Southern Great Plains, and Central Great Plains tillage management regions. Nonirrigated sugar beet (*Beta vulgaris* L.) is localized in the eastern edge of the Northern Great Plains and western edge of the Northern Corn Belt; irrigated sugar beet is also localized in the Pacific Northwest and the western portion of the Central Great Plains and the Southern Great Plains tillage management

Table 2. Climatic characteristics and winter crop potential in each of 11 tillage management regions.

Tillage Management Region*	Mean Temperature January	Monthly (°C)† July	Mean Annual Pan Evaporation† (cm)	Mean Annual Precipitation† (cm)	Frost-free Season† (days)	Growing Season Precipitation (% of annual)	Winter Crops‡
A	-9 (-14, -4)	22 (21, 23)	75 (70, 90)	76 (50, 110)	140 (120, 160)	>50	clover
B	-4 (-9, 2)	24 (23, 25)	90 (85, 100)	90 (60,110)	170 (150, 200)	50	sg clover
C	3 (0, 10)	22 (20, 27)	100 (70, 100)	120 (90, 150)	190 (120, 180)	<50 in south >50 in north	sg clovers a&b
D	5 (0, 10)	24 (20, 25)	90 (80,100)	120 (90, 150)	200 (160, 240)	<60	sg clovers a&b
E	8 (5, 10)	25 (25, 27)	105 (100, 120)	140 (130, 155)	240 (200, 280)	>60 in east <60 in west	sg clover b
F	5 (-1, 10)	27 (26, 28)	150 (120, 190)	60 (40, 100)	200 (130, 250)	>70	sg
G	-4 (-6, -1)	25 (24, 26)	90 (100, 130)	60 (40, 90)	160 (150, 170)	>70	sg
H	-12 (-17, -6)	20 (18, 24)	80 (70, 100)	40 (25, 60)	120 (100, 150)	50	none
I	0 (-5, 0)	18 (15, 20)	90 (80, 100)	35 (15, 60)	160 (100, 200)	<40	sg
J	5 (2, 10)	27 (25, 28)	110 (100, 130)	130 (130, 155)	220 (200, 260)	60	sg clover b
K	-8 (-12, -4)	18 (18, 21)	70 (60, 80)	100 (100, 115)	110 (90, 120)	60	clover a

*See figure 2 for name and letter.

†Observations show characteristic value on first line and range on second line. Source for mean annual precipitation and pan evaporation is Kohler and associates (*57*); other characteristics obtained from the Soil Conservation Service (*99*).

‡Clover A includes alsike, red, and white clovers, which are biennial or perennial in cooler climates and annuals in warm climates. Clover B includes crimson and white clovers, which have a stronger winter annual type growth. Small grains (sg) are those (wheat, rye, barley) with winter habit. Winter oats are adapted in Coastal Plains, Piedmont, and Eastern Uplands.

regions. Flax (*Linum usitatissum* L.) is exclusive to the eastern portion of the Northern Great Plains and the western edge of the Northern Corn Belt tillage management regions. Rice (*Oryza sativa* L.) is almost exclusive to the Associated Delta tillage management region. Field bean (*Phaseolus vulgarius* L.) is grown in the Pacific Northwest, Northeast,

Table 3. Distribution of crop types (planted cropland) and all cropland among tillage management regions (TMR) (23).

		Percent of Crop Area Planted in Each Tillage Management Region†							
TMR*	Percent of United States Cropland	Vegetable and Truck#	Corn	Soybean	Sorghum	Cotton	Small grain	Forages	Other§
A	18.6	24.1	47.0	33.4	0.6	—‡	8.6	18.4	7.9
B	13.1	4.5	22.8	31.3	18.5	—	6.7	12.3	1.8
C	4.7	6.3	6.3	7.8	7.8	9.7	2.7	12.3	3.0
D	1.3	3.4	2.2	2.1	1.1	0.3	1.1	2.6	1.4
E	4.2	16.4	4.3	8.7	3.4	7.6	2.8	7.5	9.6
F	15.1	5.6	2.6	0.9	41.4	60.8	27.1	14.0	19.0
G	7.8	1.4	8.2	1.1	14.4	—	11.4	8.4	10.0
H	13.1	6.2	2.4	1.6	0.3	—	29.2	10.3	25.7
I	3.7	16.9	0.5	—	—	—	7.6	6.2	9.7
J	4.9	3.6	8.4	12.6	12.5	21.5	2.0	1.0	10.8
K	1.2	11.6	2.9	0.4	—	—	0.9	7.0	1.2

*Tillage Management Region (see Figure 2).
†Total land area included is 1.3 x 10^8 hectares or 87.7% of the United States total. Sum of percentages in each column should be about 100.
‡Percentage <0.1
§Includes rice, sunflower, sugar beet, lentil, field bean, and guar (*Cyamposis* sp.).
#Includes pea and potato.

Table 4. Adoption of the various forms of conservation tillage—planting for nonforage crops in1985 in tillage management regions (TMR) with arid to dry subhumid climates in the United States (23).

TMR and Crop*	Percent of Planted Crops	Percent of Cropland Planted to Nonforage Crops						Percent of Cropland in Fallow or Idle
		NT	RiT	ST	MT	ReT	CT	
Pacific Northwest		3.6	0.1	0	20.7	7.9	32.3	32
Small grain	70	4.1	0	0	27.0	9.6	40.6	
Other†	25	2.1	0.4	0.2	2.5	3.2	8.5	
Northern Great Plains		3.6	0.1	0	16.0	9.5	29.3	32
Small grain	77	4.4	0	0	17.5	10.8	32.7	
Other‡	15	0.5	0.2	0	8.1	3.7	12.4	
Corn (FS)	5	0.9	1.3	0.1	17.1	7.2	26.6	
Central Great Plains		3.5	2.0	1.3	24.0	14.8	45.7	27
Small grain	47	1.9	0	0	23.2	18.2	43.3	
Corn (FS)	28	3.2	6.2	3.7	25.8	11.2	50.0	
Sorghum (FS)	12	11.2	0.8	0.6	30.9	13.3	56.8	
Soybean (FS)	3	2.6	3.8	3.6	21.2	15.1	46.3	
Other§	6	3.1	0	0	10.5	8.0	21.6	
Southern Great Plains		1.2	0.2	0.1	16.7	12.3	30.5	24
Small grain	56	0.8	0.1	0.1	21.0	15.1	37.1	
Sorghum (FS)	16	1.9	0.8	0	15.6	12.0	30.4	
Cotton	12	0.1	0.1	0.1	3.3	1.1	4.7	
Other#	14	2.9	0.3	0.2	12.0	10.6	26.0	

*FS represents full season; DC represents double crop.
†Corn, pea, lentil, bean, potato, and sugar beet.
‡Soybean, sugar beet, sunflower, and potato.
§Sugar beet, potato, sunflower, soybean (DC), and sorghum (DC).
#Corn, soybean, sugar beet, sunflower, and guar.

Table 5. Adoption of various forms of conservation tillage-planting for nonforage crops in 1985 in tillage management regions (TMR) with subhumid to humid climates in the northeastern United States (23).

*TMR and Crop**	*Percent of Planted Crops*	*Percent of Cropland Planted to Nonforage Crops*						*Percent of Cropland Idle*
		NT	*RiT*	*ST*	*MT*	*ReT*	*CT*	
Northern Corn Belt		3.1	1.4	0.2	31.5	0.4	36.6	5
Corn (FS)	52	4.3	1.7	0.2	31.7	0.4	38.3	
Soybean (FS)	30	1.6	1.7	0.1	33.3	0.3	37.0	
Small grain	12	1.6	0	0	33.0	0.6	35.2	
Other	3	3.2	0.2	0.2	6.4	0.5	10.4	
Southern Corn Belt		7.3	0.7	0.2	29.4	2.7	40.3	5
Soybean (FS)	38	3.6	0.6	0.1	28.4	2.0	34.7	
Corn (FS)	36	9.1	1.2	0.2	30.1	1.7	42.3	
Small grain	13	2.9	0	0	32.7	4.6	40.2	
Sorghum (FS)	7	5.4	0.4	0.3	32.8	8.4	47.4	
Soybean (DC)	3	54.4	0.1	0	19.3	5.5	79.2	
Other‡	1	16.2	0	0.2	13.6	1.3	31.3	
Northeast		6.2	0.1	0	7.0	8.9	22.2	3
Corn (FS)	50	8.0	0.2	0	7.7	11.5	27.4	
Small grain	18	3.1	0	0	4.2	5.2	12.5	
Soybean (FS)	5	0.5	0.3	0	16.5	9.3	34.6	
Other§	14	2.6	0.1	0	4.2	4.1	11.0	

*FS represents full season; DC represents double crop.
†Sugar beet, pea, sunflower, potato, and vegetable crops.
‡Corn (DC), sorghum (DC), and vegetable crops.
§Corn (DC), sugar beet, potato, pea, and vegetable crops.

Table 6. Adoption of the various forms of conservation tillage-planting for nonforage crops in 1985 in tillage management regions (TMR) with subhumid to humid climates in the southeastern United States (23).

*TMR and Crop**	*Percent of Planted Crops*	*Percent of Cropland Planted to Nonforage Crops*						*Percent of Cropland Idle*
		NT	*RiT*	*ST*	*MT*	*ReT*	*CT*	
Eastern Uplands		15.4	0	0.1	14.9	8.4	38.8	6
Corn (FS)	28	22.7	0	0.2	13.8	11.1	47.8	
Soybean (FS)	24	6.4	0	0	14.1	7.8	28.3	
Small grain	15	7.4	0	0	24.9	12.0	44.3	
(DC) crops†	9	49.9	0	0	17.1	6.3	14.3	
Other‡	20	2.8	0	0	6.8	3.4	13.0	
Piedmont		22.7	0.3	0.3	16.4	8.0	47.7	3
Corn (FS)	32	26.8	0.2	0.2	17.9	13.4	58.5	
Small grain	21	6.9	0	0	23.3	8.0	38.2	
Soybean (FS)	16	12.0	0.6	0.7	13.0	4.2	30.4	
(DC) crops	16	53.9	0	0	19.6	5.5	79.0	
Other‡	10	5.4	0.2	0	5.6	0.1	11.3	
Coastal Plains		8.6	0.1	0.6	13.1	2.8	25.2	6
Soybeans (FS)	23	4.8	0	0.3	11.2	4.0	20.2	
Corn (FS)	21	13.5	0.5	1.0	19.0	2.0	36.0	
Small grain	17	2.2	0	0	19.7	3.2	25.2	
Soybean (DC)	13	27.8	0	1.2	14.6	3.2	46.8	
Other§	22	2.1	0	0.5	3.6	1.5	7.7	
Associated Delta		2.0	0.2	0	4.4	5.3	11.7	12
Soybean (FS)	44	0.3	0	0	2.5	5.1	8.0	
Sorghum (FS)	13	0.9	0	0	3.1	5.2	9.2	
Small grain	11	8.8	0	0	11.0	8.8	28.6	
Cotton	11	0	0	0	1.0	0.4	1.4	
Soybean (DC)	5	8.0	0	0.1	10.3	19.0	37.6	
Other#	15	1.9	0.1	0	6.7	2.6	11.3	

*FS represents full season; DC represents double crop.
†80% soybean and 20% corn plus sorghum.
‡Cotton, sorghum (FS), and vegetable crops.
§Sorghum (FS and DC), corn (DC), vegatable crops, rice.
#Corn, vegetable crops, rice.

and Northern Great Plains; the areas of greatest concentration are the eastern edge of the Northern Corn Belt and western edge of the Central Great Plains tillage management regions. Lentils (*Lentilla lens* L.) are found almost exclusively in the northern portion of the Pacific Northwest tillage management region.

Among vegetable and truck crops are pea (*Pisum sativum* L.) and potatoes (*Solanum tuberosum* L.). Pea is adapted to the northern tillage management regions (Pacific Northwest, Northern Corn Belt, and Northeast), whereas potato is also found in the Northern Great Plains and Eastern Uplands tillage management regions. Among forage crops are alfalfa (*Medicago sativa* L.), clovers (*Melilotus* and *Trifolium* sp.), and perennial forage grasses. Alfalfa is widely scattered among tillage management regions, whereas clovers are found in the subhumid and humid tillage management regions. Red (*Trifolium pratense* L.), white (*Trifolium repens* L.), and alsike (*Trifolium hybridum* L.) clovers are adapted in the Northern/Southern Corn Belt, Northeast and northern portion of the Piedmont and Eastern Uplands, whereas crimson (*Trifolium incarnatum* L.) is a winter annual adapted in the Piedmont, Associated Delta, Coastal Plains, and Eastern Uplands tillage management regions (Table 2). Sweetclover (*Melilotus* sp.) is widely adapted wherever annual precipitation exceeds 40 centimeters. Martin and associates (*76*) described the adaptation of perennial forage grasses.

The crop rotation information of Daberkow and Gill (*27*) can be used to identify dominant crop rotations in the various tillage management regions. Information about crop rotation, tillage rotation, and their mutual association cannot be obtained directly from the conservation tillage survey (*23*).

Direct linkages of crop adaptation to climatic conditions are essential to the development of conservation tillage because tillage is used to modify the crop environment. Double-cropping (soybean, corn, sorghum as the summer crop) is practiced where there are more than 180 continuous frost-free days. This also defines the northern limit of crimson clover. Numerous climatic factors are suggested for corn adaptation (*96*); average night temperature during summer must be above 13 degrees celsius, warmest month isotherm must not exceed 27 degrees celsius, growing season must exceed 1,200 growing degree days, and average annual precipitation must exceed 40 centimeters. Soybean climatic requirements are similar to corn, although soybean adaptation extends further south than corn, provided that mean annual precipitation exceeds 50 centimeters. Cotton requires a frost-free season of at least 180 days, with mean summer temperature greater than 25 degrees celsius and mean annual precipitation greater than 50 centimeters, as in the Associated Delta, Coastal Plains, and eastern portion of the Southern Great Plains tillage management regions. However, cotton is grown dryland in western Texas where annual precipitation is somewhat less than 50 centimeters. Small grains, especially barley and wheat, are adapted in semiarid to subhumid climates with mean summer temperatures below about 20 degrees celsius. A frost-free period of 90 to 120 days is sufficient for small grains. Sorghum requires a mean temperature near 27 degrees celsius and is tolerant to drouth, indicative of tillage management regions in which potential evaporation exceeds mean annual precipitation. Likely winter cover crops are listed in table 2; recent research suggests new possible cover crops and their management in most of the tillage management regions (*40, 115*).

Many discussions concerning crop adaptation to climate (*76*) lack specificity. A physiologic model with stochastic weather inputs is needed to more closely characterize climatic factors affecting crop adaptation and any influences caused by tillage. Crops require significantly different moisture and thermal environments during germination and emergence, and these factors are readily modified by tillage. Numerous climatic factors are important, and their critical ranges change during the phenology of a crop; often there is a wide variation in climatic factors (thermal- and moisture-related) among years and seasons (*106*). Soil factors, including those affected by tillage, may be even more limiting than climatic factors.

Martin and associates (*76*) ranked the soil moisture factor second to climate as a determinant of crop adaptation. This factor includes capacity of the soil to store water in the root zone, but capacity also relates to the ratio of pan evaporation to annual precipitation, distribution of precipitation between growing season and dormant season for full-season crops, and mean monthly January temperature (as an index of the duration of frozen soil). Slope steepness also influences runoff and the potential supply of available water. Impeded drainage may truncate a viable root zone. Undoubtedly, these two factors play a greater role in crop adaptation at the outer fringes of the adapted region. Year-to-year and short-term fluctuations of air and soil temperature would play a greater role in crop adaptation in fringe areas. All of these factors can be modified somewhat by tillage management. Other edaphic factors, such as soil pH, salinity, and mechanically resistant layers, also influence crop adaptation (*76*).

The soil moisture factor is influenced by the actual rooting depth as related to subsoil layers. The Eastern Uplands, Piedmont, and Coastal Plains tillage management regions all have a high frequency of soil characteristics (shallow soils, fragipans, tillage pans, or acid subsoils) that reduce rooting depth and limit water availability to a root system (*19*).

Variations of soil properties have been used to predict the value of soil as a medium for root growth (*87*). This method could be used to predict crop production within a tillage management region for various crops and soil taxons as long as there are minor climatic variations. The method of Pierce and associates (*87*) integrates soil storage of available water, mechanical resistance (bulk density) to rooting, and pH suitable for rooting with depth. Climatic variations across a tillage management region would not be included, however.

Localization of Serious Erosion in the United States

Because conservation tillage is used to control soil erosion, it is informative to compare locations of serious erosion with areas of con-

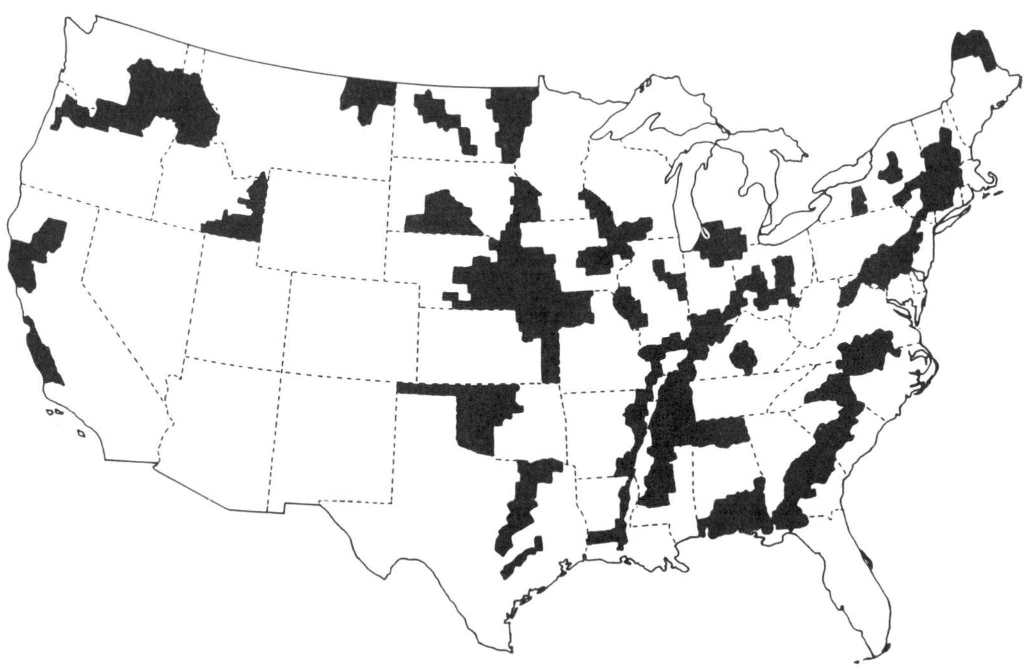

Figure 3. Distribution of serious soil erosion greater than the tolerance value in the United States, based on county wide estimates made in the 1977 National Resource Inventory (*100*).

servation tillage. The distribution of serious soil erosion (water and/or wind) is shown in figure 3; the darkened sections of the figure represent a mean annual soil erosion greater than the established tolerance value. Depending on soil characteristics, this tolerance value ranges from 4.5 to 11.2 tons per hectare per year. The erosion distributions were based on county estimates (*100*). There are serious erosional areas within each tillage management region (Figure 2), which would require conservation tillage adoption statistics to be county-based to test for greater concentrations of conservation tillage in areas with more soil erosion. Currently, this information is being collected (*21*) for highly erodible land, but the information has not been aggregated beyond the state level. Percentages of cropland in the highly erodible land category (due to wind/water erosion) suggest that figure 3 underestimates the amount of land on which conservation practices, including conservation tillage, are needed to meet the conservation compliance provision of the 1985 farm bill.

Adoption of Conservation Tillage-Planting

Overall adoption of conservation tillage[1] in 1985 ranged from a low of 11 percent of cropland planted to nonforage crops in the Associated Delta to a high of 47 percent in the Piedmont tillage management region (Tables 4, 5, 6). There were three ranges of adoption: less than 26 percent in the Associated Delta, the Northeast, and the Coastal Plains; moderate adoption of 29 to 37 percent in the Pacific Northwest, Northern Great Plains, Southern Great Plains, and Northern Corn Belt; and adoption greater than 38 percent in the Southern Corn Belt, Eastern Uplands, Central Great Plains, and Piedmont tillage management regions.

Partial-width forms of conservation tillage, with no residue disturbance from harvest until planting of the next crop, were a small part of the adopted conservation tillage-planting forms. No-till (NT) ranged from 1 to 4 percent of cropland planted in the arid to subhumid tillage management regions (Table 4), in the Associated Delta (Table 6), and in the Northern Corn Belt (Table 5). NT ranged from 6 percent in the Northeast to 23 percent in the Piedmont; adoption in the Southern Corn Belt, Eastern Uplands, and Coastal Plains tillage management regions ranged from 7 to 15 percent. The lesser use of NT in the Coastal Plains than in the adjacent Piedmont and Eastern Uplands is due to the more severe natural and traffic-induced soil compaction problems in soils of the Coastal Plains tillage management regions (*33, 61, 91*). The greatest use of NT was in conjunction with double-cropping, likely for conservation of soil moisture at planting and timeliness (speed) of planting after harvest of the first crop. Strip-till (ST) has been adopted somewhat in the Coastal Plains because of the need for a soil-shattering operation in the row in conjunction with the planter. ST has undergone some adoption in the Central Great Plains tillage management region, but the need has not been identified. Ridge-till (RiT) has been adopted somewhat for corn and soybean in the Northern/Southern Corn Belt and the Central Great Plains tillage management regions. It is used some for sorghum full season in the Central Great Plains tillage management region.

Mulch till (MT), with full-width tillage in the interim between harvest and planting, was the most-used conservation tillage form overall (Tables 4, 5, and 6). It ranged from 31 and 29 percent of cropland seeded to nonforage crops in the Northern and Southern Corn Belt, respectively, to a low of 4 and 7 percent in the Associated Delta and Northeast tillage management regions, respectively. MT ranged from 16 to 24 percent in the arid to dry subhumid tillage management regions (Table 4). Undoubtedly, mulch tillage is required in many parts of these tillage management regions to control wind and water erosion (*31*). It is interesting that NT is used about as much as MT in the Eastern Uplands, Piedmont, and Coastal Plains tillage management regions.

Adoption statistics for reduced till (ReT) are

[1]A convention in this manuscript is to use abbreviations when referring to tillage systems as defined in table 1 and to spell out the names when referring to conservation tillage systems envisioned in the abstract definition.

difficult to interpret. However, it should be considered as full-width tillage in which a primary tillage operation is combined with the seeding operation in a one-pass operation to avoid secondary tillage. Because of the potential for more weed and volunteer growth and use of a cover crop during the winters in mid-latitude to southern tillage management regions, ReT is used most frequently for summer crops in the Central Great Plains, Southern Great Plains, Piedmont, and Eastern Uplands tillage management regions. There is unusually low use of ReT in the Coastal Plains, and Northern and Southern Corn Belt tillage management regions. Moderate to high use of ReT in the arid to dry subhumid tillage management regions (Table 4), higher use of ReT for small grain, and the dominance of small grain suggests that ReT may be used frequently for seeding small grain into stubble of a preceding small grain.

Within the various tillage management regions, adoption of conservation tillage for the various crops is quite different (Tables 4, 5, and 6). These comparisons indicate that crop rotation (and sometimes also tillage rotation) is a necessary component of conservation tillage. Overall, conservation tillage is highest in those tillage management regions where the percentage of planted crops (Table 3) is not dominated by one or two crops.

The two northern-most tillage management regions and the subhumid to arid tillage management regions are examples in which less adoption takes place because only a few crops are adopted. Summer fallow and small grain constitute 78 percent of the cropland in the Pacific Northwest; land available for other full-season adopted crops, such as lentils and peas, severely limit crop rotation except in small areas of the tillage management region. Small grain monoculture or even a 2-year rotation (wheat-summerfallow) impedes MT and ReT because the associated surface residue and root tissue (even in small quantities) often harbors inoculum for fungal-type root diseases (*25*).

A somewhat similar situation occurs in the Northern Great Plains tillage management region where summerfallow and small grain account for about 84 percent of the cropland. The disproportionately high production of small grain in this tillage management region indicates small-grain-after-small-grain sequences, which encourages crop residue burial to avoid disease (*25*).

Crop residue production in the Northern Great Plains tillage management region is also low enough to restrict secondary tillage consistent with achieving a 30 percent cover after planting. Although conservation tillage for other crops in the tillage management regions of table 4 is low (peas and lentils in the Pacific Northwest and sunflower in the Northern Great Plains), these crops are valuable for a conservation tillage system with small grains because they reduce the build-up of pathogen inoculum associated with monoculture (*24*). There is recent evidence that pea (*121*) and sunflower (*28*) produce well in tillage systems in which sur-

Table 7. Adoption of conservation tillage-planting forms for forage crops in 1985 in each of the 11 tillage management regions (TMR) (23).

*TMR**	*Percent of Planted Crop*	*Percent of Cropland Planted to Forages†*			
		NT	*MT*	*ReT*	*CT*
A	2.2	2.1	23.9	0.1	26.2
B	2.3	7.0	15.9	1.7	24.6
C	6.5	12.0	10.9	5.7	28.7
D	4.7	18.5	12.6	5.6	36.7
E	4.2	5.0	8.0	4.7	17.8
F	2.7	1.7	9.2	9.1	20.1
G	3.3	3.6	22.1	10.0	35.7
H	2.5	1.0	6.0	3.4	10.4
I	5.4	1.6	1.2	4.6	7.4
J	0.5	9.9	6.9	9.5	26.2
K	13.6	3.6	0.9	1.5	6.0

*See figure 2 for name of TMR.
†No significant use of ridge-till (RiT) or strip-till (ST) to plant forage crops.

face cover with crop residue may be less than 30 percent, yet the residue is buried within 10 centimeters of the surface.

High adoption of conservation tillage in the Central Great Plains results from a large selection of crops and high use of all forms of conservation tillage, at least for some crops (Table 4). A common rotation is wheat-grain sorghum (corn)-summer fallow (about 20 percent of the sequences involving wheat) (*27*), with high use of mulch tillage and especially no-till and strip-till for full-season sorghum and full-season corn planted into the winter wheat stubble (*89, 118*). No-till planting of corn or sorghum into winter wheat stubble is termed ecofallow (*119*) because herbicide control of weeds and no tillage produce a high recharge of soil water during fall and winter. Small grain-soybean rotations (about 10 percent of the crop rotations involving soybean) (*27*) with the same ecofallow management of weeds in small grain is practiced in the eastern portion of the Central Great Plains tillage management region (*14*).

A high percentage of ReT for planting of all four major crops suggests flexible use of conservation tillage systems in these crop rotations. The high use of RiT for full-season corn and the inability to mix RiT with other forms of tillage in a crop rotation suggests that there is much monoculture corn in RiT systems. However, the high use of RiT for soybean in this tillage management region suggests the beginning of corn-soybean rotation in RiT. Monoculture production of corn and sorghum with ridge tillage had already been achieved in this tillage management region about 1970 (*125*).

At least three distinct cropping systems can be found in the Southern Great Plains (*54*): (a) grain sorghum-summerfallow-wheat in the northern and western subregions, (b) nearly a monoculture (3 out of 4 years) of winter wheat in the northern and eastern subregions, and (c) an area of cotton monoculture with very little use of conservation tillage (Table 4). In Texas, Oklahoma, and Kansas, the frequency of wheat after wheat was 54, 81, and 28 percent, respectively, for wheat planted in 1988 (*27*). Cotton in a monoculture or after fallow constitutes 70 percent of the cotton planted in Texas (*27*). MT and ReT are used equally as frequent for rotations consisting of small grain-grain sorghum-summerfallow. MT is not used in the subregion with monoculture winter wheat, but ReT is facilitated because of the large amount of weed and volunteer crop growth and the low use of herbicide (*34*) in this cropping system. There is an interesting use of NT and RiT for grain sorghum beginning in this tillage management region (*54*).

The most dominant crop rotation in the two Corn Belt tillage management regions is corn-soybean, while a rotation of corn-soybean-small grain is used much less frequently (*27*). Rotations of corn-corn-soybean are also used in the Northern Corn Belt; continuous corn constitutes about 25 percent of the corn planted in the two Corn Belt tillage management regions, but it constitutes nearly 35 percent in states of the Northern Corn Belt tillage management region (*27*). Monoculture or uninterrupted succeeding crops of soybean are infrequent and are hazardous due to diseases of *Phytophthora* spp. and *Cephalosporium gregatium*, especially the former if soil drainage is poor and conservation tillage is used. Soybean cyst nematode (*Heterodera glycines*) is also a serious hazard in soybean monoculture. Corn is a more frequent crop than soybean in the Northern Corn Belt, whereas the frequency of corn and soybean is nearly the same in the Southern Corn Belt; grain sorghum is also a significant crop in the Southern Corn Belt tillage management region (Table 5). MT is the major conservation tillage in these tillage management regions, and more corn than soybean is planted NT (Table 5) because, in a rotation of the two crops, NT would be more difficult for planting soybean in corn residue. Perhaps MT is preferred for soybean after corn because it responds better to narrow rows (*50*) and the facility of narrow rows in MT compared with NT. MT or NT may be used in monoculture or succeeding years of corn, especially in the Northern Corn Belt tillage management region.

Similar areas of corn and soybean planted with RiT in the Northern Corn Belt suggest a corn-soybean rotation, but the imbalance in the Southern Corn Belt tillage management region suggests some continuous corn or corn in rotation with another crop, not RiT planted, such as close-seeded soybean. Forcella and Lindstrom (*38*) demonstrated relatively more difficult weed control in RiT with continuous corn than when rotated with soybean. The movement of soil and weed seeds off the ridge before planting in the ridge tillage system demonstrates a cultural weed control advantage of ridge tillage over conventional tillage (*39*). Because soil temperature in the row area can be increased when crop residue is cleaned from the row in the spring (*105*), ridge tillage is a desired conservation tillage type for corn in the northern latitudes. The shoot meristem of corn emerges above ground when the sixth leaf emerges about 20 days after shoot emergence (*66*). Until this event, soil temperature, not air temperature, controls corn growth. Soybean is less sensitive to soil temperature because the dicotyledon growth habit thrusts the growing point of the shoot above the soil surface coincident with emergence (*68*). Ridge tillage has been used in the two Corn Belt tillage management regions for corn monoculture since about 1965 (*42, 45*), but its use for a soybean-corn rotation began about 1970 (*35, 90*). There is need to include a close-seeded crop (small grain or an annual food legume) in a corn-soybean rotation without destroying the ridge design in the field; a major benefit would be cultural control of disease and insect damage to roots of these crops.

In the northern tillage management regions, with a subhumid to humid moisture regime (Northern Great Plains, Northern Corn Belt, Northeast), surface residues may reduce soil temperature in the spring. As discussed above, corn growth may be retarded by reduced soil temperature, whereas the dicotyledon growth habit of soybean makes it sensitive to air temperature. Small grains have a thermal sensitivity somewhat less than that of corn; temperature affects the rate of tiller production (*6*). In a corn-soybean rotation, avoidance of soil temperature problems on corn is not difficult with conservation tillage as long as the row is cleared (*105*). However, conservation tillage is difficult for soybean planting because corn residue amounts are at least 200 percent greater than soybean residue amounts, and the residue is more resistant to decomposition. Bull (*12*) noted this problem with respect to frequency of moldboard plowing (e.g., the use of moldboard plow relative to a chisel or disk for primary tillage in the Corn Belt states was much higher for land preparation after corn than after soybean). The higher use of no-till for corn planting than for soybean planting (*12*) is another reflection of the same problem. The emergence of soybean is extremely sensitive to variations in planting depth and nonuniform packing around the seed, which can occur when large amounts of corn residue are near and on the surface (*50*).

A typical double-cropping system is a winter small grain, harvested or grazed, with soybean no-till seeded into the standing residue (*49, 61, 69, 93*). Sorghum double-cropping or corn double-cropping is most likely used when the winter cover crop is a legume. Double-cropping (mainly soybean, but some corn and grain sorghum as the summer crop) has been adopted somewhat in the Southern Corn Belt tillage management region, with roughly 55 percent of the soybean planted as a double crop by NT (Table 5). Some double-cropping of soybean and sorghum was reported in the Central Great Plains tillage management region. Double-cropped soybean as a percentage of the planted crop is highest in the Piedmont, Eastern Uplands, Coastal Plains, and Associated Delta tillage management regions (Table 6). Marra and Carlson (*75*) estimate that double-cropped soybean constitutes 30 to 40 percent of the soybean planted in the tillage management regions of the southeastern United States. Sanford and associates (*94*) estimate that about 80 percent of the double-cropped soybean is in a rotation with small grain in this geographic region.

The dominant conservation tillage methods for planting soybean are NT and ReT, with some use of ST in the Coastal Plains because of the need for an in-row chisel ahead of the planter opener to relieve excessive soil mechanical resistance due to soil compaction. Conservation tillage-planting of small grain appears to be MT when in the double-cropping rotation. For nearly all double-cropped soybean in rotation with small grain, small grain is mulch tilled or conventionally tilled to avoid soil compaction and poor drainage of Oxisols and Alfisols during winter (*46, 64*). Wesley and Cook (*117*) demonstrated that monetary returns from wheat encouraged a double-cropping system of no-till for soybean and a disk-based tillage for wheat planted on poorly drained Inceptisols typically found in the Associated Delta tillage management region. However, grain stubble burning enhanced yields of no-till-planted soybean. The high use of NT for full-season corn in the Piedmont (southern portion), Eastern Uplands, and Coastal Plains tillage management regions is a popular system (*16, 17, 82, 86*) involving no-till planting into a winter cover crop (legume, small grain, grass). In all tillage management regions listed in table 6, full-season soybean has a lower use of CT than double-cropped soybean and full-season corn. Full-season soybean grown in rotation with small grains, full-season corn, or full-season sorghum require a seedbed that often is difficult to obtain with the large amounts of surface residues produced by these crops, especially if the row spacing is close enough to prevent clearing the surface residue away from the row and if the surface residue from these crops is not partially incorporated in the previous fall.

Except for in the Associated Delta, the overall use of CT is much lower for forage than for nonforage crops (Table 7 versus Tables 4, 5, and 6) and bears no relation to the amount of cropland seeded to forage crops. The greatest use of CT for forages is in the Central Great Plains and Piedmont tillage management regions in which there was the highest use of CT for nonforage crops. NT planting of forage crops appears to be greater in the humid and southern tillage management regions. Except in very special circumstances, a conservation tillage system can use forage crops for the necessary crop residue into which nonforage crops (usually warm-season types) can be planted with a minimum of surface residue disturbance. If the cover crop is planted and the seedlings established during a period of low soil erosion, there is less need for conservation tillage-planting of the forage/cover crop.

For instance, Mannering and associates (*74*) suggest that wheat, rye, or hairy vetch (*Vicia villosa* L.) be seeded as a cover crop in Indiana following corn/soybean. Seed broadcast without soil disturbance before harvest is considered risky; the recommended procedure is to lightly disk, fertilize, and seed. This full-width tillage helps avoid problems of herbicide carryover, drill penetration for seeding close-seeded crops, and low pH zones near the surface. The full-width fall tillage also prepares a good seedbed for no-tillage or strip-tillage planting of the next summer crop.

Factors Affecting Adoption of Conservation Tillage

Within the last 10 years, there has been a convincing argument that economic feasibility is the cardinal factor determining adoption. Yet, conservation tillage was designated by the U.S. Department of Agriculture in 1980 as "the most economical and effective means of soil erosion control." New technology needed for conservation tillage-planting systems often ranks high as an impeding factor. The conservation benefits of conservation tillage are recognized, but are not a major factor in adoption. Several surveys can be cited to amplify these conclusions.

A survey (*22*) of growers active in the policy of soil conservation districts listed the following seven reasons for adoption (in decreasing order of importance): (greater) profit, fuel savings, fewer trips (over the field), (more) erosion control, save soil moisture, reduce compaction, and reduce sediment (loss). The first three reasons are economic and/or cost-oriented, and the remaining four are resource conservation-related. The least important of additional reasons were pest control and ease of management.

Farmers from all parts of the United States agreed on six obstacles to adoption of conservation tillage (in decreasing order of importance): weed control, finances, grower attitude (or reluctance), inadequate equipment, poor net profit, and difficult crop residue management. The farmer's list of needs nearly paralleled the obstacles; they also included disease and insect control, fertilizer management, and adapted plant varieties. Clearly, adoption was more sensitive to economic rather than resource conservation factors. A wide range of technological deficiencies also impeded adoption and raised the risk of failure. Significant improvements in weed control (*120*) as well as in tillage and planting machinery systems (*36, 51, 81*) have been made even since 1980. Significant among these machinery advancements is the handling of crop residue, precise placement of fertilizers, and depth control for seed delivery.

A farm survey in Iowa (*116*) sought to determine the degree of advantage or disadvantage of leaving at least one-third of the previous year's corn residue on the soil surface at soybean planting. Responses echoed those discussed above, in that technical factors (pest control, planting problems, emergence uniformity, and fertilizer application) were most frequently a disadvantage, whereas economic factors of profit, fuel use, and operator time were most frequently considered an advantage. Cost of machinery conversion, however, ranked high as a disadvantage. The value of conservation tillage and surface residue for resource conservation (erosion control and water conservation) were clearly recognized as most advantageous.

A survey in Ohio (*59*) found many points of agreement with the Iowa survey (*116*). The Ohio survey (*59*) did find that many farmers were aware that specific soils were not suitable for conservation tillage or had unsuccessful experiences with conservation tillage, including no-till.

Most economic analyses have projected that one or more conservation tillage systems will provide larger net returns than conventional tillage. In these analyses, we have considered conservation tillage as the use of a chisel, disk, or sweep for primary tillage even though there may be less than 30 percent of soil surface covered by residue after planting. However, there are instances in which the projected net return from conservation tillage is smaller than from conventional tillage.

Most net return analyses evaluate tillage systems for corn and soybean production in the Corn Belt. Larger net returns were projected for corn production with conservation tillage systems (primary tillage with a chisel or disk), relative to a moldboard-plow based system, on a poorly drained soil in Iowa (*9*) and for both soybean and corn production on a moderately well-drained soil in Iowa (*52*). For soybean production in a corn-soybean rotation, Brown and associates (*9*) found equal net returns for conservation and conventional systems. In the systems studied by Brown and associates (*9*), the surface covered by residue averaged 72, 26, and 3 percent for the no-tillage, reduced tillage (chisel- or disk-based), and conventional tillage (moldboard-based) systems, respectively. Klemme (*58*) found equal net returns from moldboard- and chisel-based systems and ridge tillage used in the production of soybean and corn on a poorly drained soil in Indiana, but the no-tillage system had lower net returns, primarily because of lower yield. Doster and associates (*30*) used relative yield ratings and costs to show that ridge-till produced larger net returns for all three groups of soils (mainly a drainage classification) in Indiana. Ridge-till and no-till produced equal returns on well-drained and sloping land, whereas on poorly drained soils net returns were nearly equal for a chisel-based full-width system and a ridge-till system. These net return comparisons in Indiana are linked with an extensive plan to recommend tillage systems based upon soil properties, topography, and climate (*42, 73*). Duffy and Hanthorn (*32*) showed no significant differences of net return among tillage systems for corn production, but found larger net returns with conventional tillage for soybean (full season) production in the Corn Belt, even though costs were lower for both corn and soybean in reduced (not moldboard-based) systems and no-till.

Net economic returns in the Central and Southern Great Plains reveal that increased crop yields enhanced the adoption of conservation tillage systems (*47, 123*). In a sorghum-soybean rotation in northeastern Kansas, Williams and associates (*47*) demonstrated that no-till, with winter ecofallow, produced higher net returns than the conventional system (disk-based) for those not adverse to risk due to year-to-year variation in soybean yield. When the operator feared risk, the largest net return was a no-till sorghum or a conventional (disk-based) system.

Highest net returns were realized in western Kansas by a conventional system (sweep- and disk-based) for production in a wheat-sorghum-fallow rotation (*123*). Irrespective of risk attitude, a no-till system in this rotation was not preferred. Net return increases lead to continuous wheat or sorghum with no-till. It was also shown that no-till ecofallow after wheat yielded larger net returns; it happens that such tillage management has approached the conventional status (*122*).

A most unusual larger net return for conservation tillage was facilitated by no-till winter fallow for continuous cotton in Texas (*56*). A no-till winter fallow without a crop yielded larger net returns because of lower costs from cotton harvest to cotton planting in the spring (cotton yield was not changed). The no-till winter fallow planted to wheat produced larger net returns because of consistent cotton yield increases.

Mulch Tillage Systems

Identity of the primary tillage tool and amount of surface covered with crop residue after planting are both needed in mulch tillage to specify the depth of incorporated residue and subsequent value for erosion control and water conservation. The value of surface cover for erosion control is also identified. Timing of the primary tillage (fall or spring) has a major impact on wind and water erosion in those tillage management regions with a cool dormant season between full-season crops.

The adoption of mulch tillage (as described in table 1) for the Northern Corn Belt, Southern Corn Belt, Northern Great Plains, and Central Great Plains tillage management regions was 31.5, 29.4, 16.0, and 24.0 percent, respectively, of cropland planted to nonforage crops (Tables 4 and 5). The use of mulch tillage to achieve 30 percent cover was not variable among crops, except for sugar beet, potato, sunflower, vegetable crops, and soybean. This amount of surface cover at planting cannot be achieved when a moldboard plow is used for primary tillage, except in rare cases (*29, 114*). Except when preceded by high residue crops, such as corn, sorghum, and small grain, a 30 percent surface cover with residue cannot be achieved at planting unless there is a stringent restriction on secondary tillage after primary tillage with a chisel, disk, or sweep. Dickey and associates (*29*) found that a surface residue cover of 15 to 20 percent measured in a Nebraska survey corresponded to the rate of MT (30 percent of surface covered with residue) adoption found in the CTIC survey of the same geographic area. Brown and associates (*9*) found that, when the number of residue-reducing secondary operations (including planting) was three or fewer after a disk- or chisel-type primary tillage, a residue cover of 20 percent could be achieved in a corn-soybean sequence. Yet Dickey and associates (*29*) found that the mean number of secondary operations ranged from 2.6 to 4.3, depending on the type of tillage tool used for primary tillage. Those with a field cultivator system for primary tillage used the least number of secondary operations and those with a moldboard plow system used the largest number of operations.

Chisels, sweeps, or disks are used for primary tillage in the Northern Corn Belt, Southern Corn Belt, Northern Great Plains, and Central Great Plains tillage management regions much more frequently than inferred by the estimated adoption of MT on 27 to 32 percent (Table 4) of planted cropland. Moreover, major shifts toward the use of these three primary tillage tools has occurred since 1980. Dickey and associates (*29*) found in Nebraska that a chisel, disk, and moldboard plow was used for primary tillage with a frequency of 29, 36, and 16 percent, respectively, in 1984. Respective uses in 1974 were 13, 29, and 51 percent. In a 1983 survey of the Corn Belt, Fletcher (*37*) noted that 38 and 52 percent of the primary tillage was done with moldboard plow and chisel plow (or disk), respectively, and projected a decline in the use of moldboard plowing. In one Iowa county in 1983 (*116*), a moldboard plow was used 56, 6, and 60 percent of the time, respectively, for tillage after corn in corn-corn sequences, after soybean in soybean-corn sequences, and after corn in corn-soybean sequences. Based upon remaining residue, a nonmoldboard tillage tool was used 44, 94, and 40 percent of the time for primary tillage in the respective sequences.

A national survey of tillage systems for planted major crops in 1987 and 1988 (*11, 12*) agreed with the limited-area surveys discussed above. A database of crop-residue yield, primary and secondary tillage tools used, crop residue burial for each tillage tool, type of residue, residue-weight-to-surface-cover algorithms, and the required 30 percent cover for a conservation tillage system were systematically used to project the relative frequency of four tillage systems: (a) conventional with a moldboard plow as the primary tillage, (b) conventional without a moldboard plow for primary tillage, (c) mulch-till, and (d) no-till. The two latter systems are considered conservational when judged by the rule of 30 percent cover. In those states of the Corn Belt tillage management regions where the corn-soybean rotation predominates, about 50 percent of the corn and soybean was planted using a conventional system without moldboard plowing and 15 percent using a conventional system with moldboard plowing. Use of a conventional system with moldboard primary tillage was about 8 percent greater for soybean (after corn) than for corn (after soybean). Mulch-till and no-till combined were used on about 15 to 20 percent of the planted soybean and corn. A similar preferred use of disk, chisel, or sweep implements for primary tillage in conventional systems was shown in the winter and spring wheat planted in the states of the Northern and Central Great Plains tillage management regions.

The much lower frequency of mulch-till (as determined by the CTIC survey, Tables 4 and 5), compared with the above observed use of sweeps, disks, and chisels for primary tillage, indicates that there is sufficient secondary tillage to markedly reduce the surface residue cover produced with these tillage tools. Johnson (*51*) showed that surface cover percentage produced by sweeps and chisels of various types should be about 50 percent after corn and 25 percent after soybean.

The type of tillage tool used for primary tillage has a marked impact on the depth of buried crop residue (Figure 4). The location of wheat residue was measured after at least two secondary tillage treatments (10 to 15 centimeters deep) in spring following fall primary tillage. The buried residue positions (Figure 4) would apply at time of planting and would have accounted for any tendency of crop residue near the surface to decay more quickly than the deeper buried residue. In no instance was there as much as 10 percent of the surface covered with residue when these measurements were made. Moldboard plow penetration ranged from 15 to 34 centimeters, chisels penetrated to 20 centimeters, and sweep penetration ranged from 13 to 19 centimeters. Sweep, chisel, and moldboard tillage buried 10, 20, and 60 percent, respectively, of wheat residue to a depth greater than 10 centimeters. Measurements in Minnesota have shown residue burial patterns similar to that presented in figure 4 (J. Staricka, 1989, personal communication). Moreover, the residue incorporated with a disk in the Minnesota study was only about 2 centimeters deeper than that with a chisel. Colvin and associates (*20*) suggest that secondary tillage with a field cultivator (sweeps

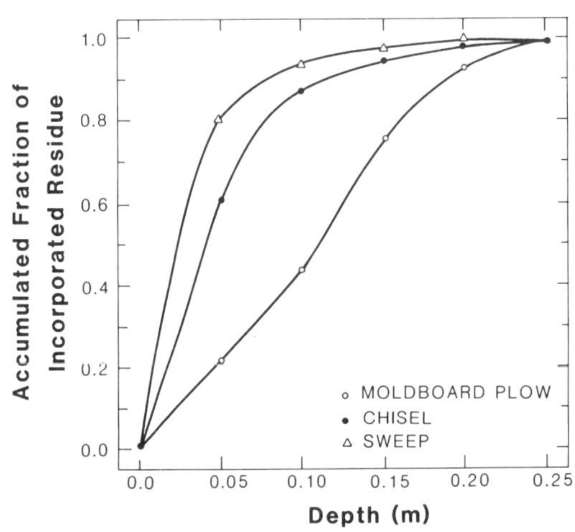

Figure 4. Buried position of wheat residue as produced by primary tillage with sweep, chisel, or moldboard plow [after Allmaras and associates (2)].

Table 8. Components of the C factor (water erosion) as related to full-width tillage used to manage crop residue and soil surface roughness.

Previous Crop	Primary Tillage and (Secondary Tillages)*	Residue Distribution after Tillage† Surface	1 to 10 cm	Below 10 cm	Fractional Surface Cover‡	Surface Random Roughness (cm)‡	C Factor and Subfactors SR	RC	PLU	C Factor
		kg/ha								
Corn	None	8,500	0	0	0.97	0.6	1.00	0.03	0.65	0.02
	MP	680	3,670	4,150	.24	3.0	.62	.47	0.61	.18
	MP (3)	100	3,280	4,150	.04	0.8	.96	.87	0.65	.54
	CP	5,100	2,700	700	.87	1.6	.78	.06	0.70	.03
	CP (2)	1,020	5,220	700	.43	1.2	.87	.24	0.50	.10
Wheat	None	6,000	0	0	0.95	0.6	1.00	.04	0.65	.03
	MP	480	2,200	3,320	0.21	3.0	.68	.51	0.75	.26
	MP (3)	50	1,830	3,320	0.02	0.8	.98	.93	0.78	.71
	CP	3,600	1,920	480	0.83	1.6	.80	.07	0.77	.04
	CP (2)	630	3,230	480	0.27	1.2	.89	.40	0.65	.23
Soybean	None	3,600	0	0	0.86	0.6	1.00	0.05	0.65	.03
	MP	290	920	2,390	0.15	2.8	.82	.61	0.88	.44
	MP (3)	10	850	2,390	0.01	0.6	1.00	.97	0.89	.86
	CP	2,160	1,150	290	0.69	1.4	.87	.10	0.86	.07
	CP (2)	90	2,410	290	0.05	1.0	.94	.84	0.73	.58

*MP represents moldboard plow and CP represents chisel plow and are assumed to be fall tillage; secondary tillages were assumed to be performed prior to spring planting; three secondary tillages assumed after moldboard plowing, two assumed after chisel plowing; secondary tillage implements were tandem disc or field cultivator.

†Moldboard and chisel plow assumed to bury 92 and 40% of the residue, respectively. Overwinter decay of corn, wheat, and soybean residues assumed to be 20, 30, and 30% respectively, of that on the surface and within the upper 10 cm. Each secondary tillage pass was assumed to reduce surface residue 50% for corn and wheat, and 75% for soybean, and this residue pool was added to that buried in the 1- to 10-cm depth. Moldboard and chisel plowing distributed 40 and 80% of the buried residue above 10 cm.

‡Surface cover computed using method of Gregory (*44*); random roughness estimated from Voorhees and associates (*114*).

§Computed using Laflen and associates (*60*); SR is related to random roughness of the soil surface; RC relates to crop residue cover; and PLU is related to crop residue buried in the 1- to 10-cm depth.

about 25 centimeters wide) or a tandem disk reduces surface residue cover (corn, wheat, or soybean residue) from 20 to 50 percent per pass depending on the type of residue.

A comparison of C factors, which depicts tillage and crop residue management for control of water erosion, demonstrates that the three subfactors of random roughness, residue cover on the soil surface, and shallow buried residue have complimentary significance (Table 8). A recently tilled, bare soil without roughness has a value of 1.0 for all three subfactors and the C factor. If a 60 percent reduction in C can be considered adequate for erosion control during the fall through preplant period, then both the chisel and moldboard plow primary tillage treatments are predicted to be adequate. However, chisel tillage would be superior because of the combined surface residue cover and shallow-buried residue. When moldboard plowing was followed by three secondary tillage treatments typical of the Corn Belt and adjacent Great Plains (*29*), predicted erosion reduction did not exceed 46 percent, whereas chisel plowing typically followed by two secondary tillages (*29*) was predicted to achieve an erosion reduction greater than 42 percent for all three residue forms (Table 8). Only after corn and wheat did the chisel system approach 30 percent surface cover, assuming that the planter had only a minor influence on surface residue cover.

Several additional features of table 8 are worthy of comment. The 50 and 75 percent reduction in surface residue cover for each secondary tillage after corn and soybean, respectively, is undoubtedly excessive. However, the estimates provide a conservatively high C-factor value because surface cover is a relatively more effective erosion control subfactor than buried residue, based upon the residue-weight concentration expressed as kilograms per hectare per millimeter of soil. Crop residue buried below 10 centimeters in the moldboard plow treatment would have no impact on erosion control during the current growing season. Soil-engaging tools on a chisel plow provide a wide variation of control over residue positioning (*51*); low to medium crown sweeps would have increased surface cover (Table 8). A disk either in a combination implement or as the only soil-engaging tool would reduce the residue cover on the surface (*20, 51*). The rill erosion reduction provided by shallow buried residue (Table 8) is a conservative estimate of the greater value of this subfactor for erosion control as found by Brown and associates (*10*) on a soil sensitive to rill erosion.

Other probable reasons for increased farm adoption of the chisel plow and disk for primary tillage are wind erosion control, enhanced water conservation, reduced input costs, and facility of pesticide application. Wind erosion can be reduced 60 percent when 20 percent of the soil surface is covered with anchored crop residue or nonerodible clods (*41*). The percentage of nonerodible clods (>0.84 millimeter diameter) in an air-dry soil mulch, which is inversely related to soil erodibility, increased by 5 to 8 percent per 1,000 kilograms wheat straw used with tillage systems that maintained the straw on the surface or buried it less than 10 centimeters (*8, 98*). Tillage tools may provide from 10 to 75 percent nonerodible clods in a soil mulch, whereas moldboard plowing may provide the highest percentage of nonerodible clods initially. Those tillage treatments that maintained residues near the surface ultimately produced soil mulches with a higher percentage of nonerodible clods (*97*). These trends with nonerodible clods parallel those with random roughness, but quantitative links are yet to be established. Shallow-buried residues undoubtedly influence soil crust susceptibility to wind erosion, as suggested by an increase in soil surface cover caused during over-winter slaking of the soil (*51*) and the large reduction in wind erosion as soil cover with anchored residue increases (*41*).

Unger and associates (*109*) cited numerous experiments showing that soil water storage decreases in the following order: surface residue, shallow-buried residue, and residue buried below 10 centimeters as with moldboard plowing. Both infiltration and evaporation are operative in this scenario.

In an earlier section, it was demonstrated that the chisel-based system is often more cost-effective than the moldboard-based system because of fewer operations; in some instances yields were also higher. It is quite possible that a chisel- or disk-based system is used more because it facilitates preplant application of herbicides. Dickey and associates (*29*) and Bull (*12*) note that there are fewer secondary tillage treatments in a chisel/disk system than in a moldboard-based system, and it is possible that the first secondary tillage after moldboard plowing is needed to reduce surface roughness sufficient to achieve a uniform depth of herbicide application. Johnson (*51*) found that chisel-based systems in the fall, especially those with sweeps, usually had soil height extremes in the spring that were less than plus or minus 50 millimeters—a limit suggested to maintain herbicides within the 25 to 50 millimeters depth. The greater random roughness after moldboard plowing versus chiseling (*114*) suggests a preference for chisel- or disk-based systems.

Tillage Systems and Soil Taxonomy: Yield Effects

Risk associated with adoption of conservation tillage is perhaps most sensitive to the year-to-year consistency of comparative crop yields, although relative input costs dominate when yields show little sensitivity to tillage method. Soil drainage suitability for the various forms of conservation tillage, which has been studied since the early 1960s, has a major influence on soybean and corn yields.

Van Doren and Reicosky (*112*) evaluated published literature (since about 1960) to systematize current knowledge about soybean response to soil properties and tillage systems. That information is provided with similar evaluations for corn grain yield (Table 9). Included are only those field comparisons in which there was equal stand and adequate weed control among tillage systems in an individual

Table 9. Effects of soil properties, cropping practice, and tillage on soybean and corn grain yield [soybean data from Van Doren and Reicosky (*113*)].

Soil Property		Cropping	Moldboard Plow§		Chisel§				Disc or Field		No-tillage§	
					Normal		Deep					
*Drainage**	*Texture†‡*	*Sequence‡*	*Fall*	*Spring*	*Fall*	*Spring*	*Fall*	*Spring*	*Cultivate§*	*Ridge§*	*<6 year*	*≥6 year*
			Soybean Grain Yield (kg/ha)									
Poor	Fine	Continuous	*3,440*	-20	-110	—	0	—	-170	—	-380	-290
		Rotation	*2,980*	-30	-70	-120	—	-30	-140	—	-230	-420
	Medium	All	-120	*2,870*	+100	-190	+20	-100	-170	—	+30	—
	Fine	DC	—	*1,380*	—	—	—	—	+360	—	+210	—
Imperfect	All	All	*2,810*	—	0	+200	—	+160	+100	—	+60	—
	Fine	DC	—	*1,500*	—	—	—	—	—	—	+80	—
Well-1	Medium	All	+320	*2,190*	+260	+80	—	+70	+20	—	+40	+240
Well-2	Coarse	All	—	*2,230*	—	+320	—	+380	+20	—	—	—
			Corn Grain Yield (kg/ha)									
Poor	Fine	Continuous	*8,020*	+80	-470	-940	-320	—	-510	-290	-650	-1,020
		Rotation	*8,120*	-230	-220	-350	+20	—	-190	-250	-310	-410
	Medium	All	*8,340*	-280	-360	+50	-10	-180	-300	-470	-710	-650
Imperfect	All	Continuous	*7,510*	-350	-210	+40	-200	—	-140	-440	-360	—
		Rotation	*7,700*	-670	+30	+200	+490	—	-90	-60	-100	—
Well-1	All	Continuous	+1,090	*6,750*	+1110	+1990	—	+870	-970	+870	+650	+1700
		Rotation	—	*6,810*	—	—	—	+890	+80	+700	+760	+1210
Well-2	Coarse	All	+260	*6,120*	—	+1080	—	—	+20	+480	+780	+1470

*Poor drainage are soils in the aquic suborder (*103*); imperfect drainage are soils from aquic and nonaquic subgroups; well-1 drainage are soils from nonaquic subgroups except Paleudults; and well-2 are Paleudults.
†Fine textures are surface horizons of silty clay loam, clay loam, or clay; medium textures are from sandy to silt loam weighted toward silt loam; coarse textures are from sandy loam to silt loam weighted toward the sandy loam.
‡All indicates insufficient data for separating texture or cropping systems into separate categories. DC is double cropping of soybean with wheat.
§Moldboard plow depth = 15 to 25 cm, normal chisel depth = 15 to 20 cm, deep chisel depth = 30 to 50 cm, disc or field cultivate depth = 8 to 12 cm, number of years consecutive no-tillage on the same plot; the italicized yield is the base yield [see Van Doren and Reicosky (*113*) for method to interpret statistically] from which all others are compared either as an increase (+) or decrease (-).

field comparison. Previous crop, yields for individual years, and physical location of the site were also required. Crop yields from various tillage systems were grouped by similar drainage, surface soil texture, and previous crop (Table 9). Poorly drained soils usually had little surface drainage and a slow internal drainage, while imperfectly drained soils usually had better surface drainage combined with slow internal drainage. Well-drained soils had good surface and internal drainage.

As soil drainage was improved, yields of both corn and soybean grown with conservation tillage treatments improved relative to moldboard plow as the primary tillage (Table 9). Moldboard plow was superior on poorly drained soils, whereas in well-drained soils nearly all conservation tillage treatments (abstract definition) were superior to moldboard tillage. On less well-drained soils, various forms of conservation tillage produced less depressed to higher corn yields relative to moldboard plowing when grown in rotation. The longer a field experiment continued, the greater the differences between corn yields in a moldboard-plow system compared with no-tillage (e.g., no-tillage yields became relatively worse on poorly drained soils and relatively better on well-drained soils). Results for corn on Paleudults (Table 9) were not significantly different than for other well-drained soils, whereas for soybean, chisel tillage produced relatively greater yields on Paleudults than on other well-drained soils. Among other soil environmental changes brought about by moldboard plowing versus chisel plowing is the almost universal fragmentation of the 15- to 25-centimeter layer with the moldboard, but not with the chisel (*2*).

Some of the corn yield differences between

moldboard plowing and no-tillage can be explained by the K value of soil in the USLE (*124*). A greater K value implies greater soil susceptibility to detachment; soil particles are more easily dislodged and available to plug pores during infiltration. If surface residue exceeded 30 percent cover, no-tillage yields increased compared to moldboard-plow yields as K increased (not shown). Thus, residue cover is more effective as the soil more easily yields dislodged particles.

These observations (Table 9) suggest that, if crop yield is a criterion, conservation tillage adoption should be proceeding on imperfect to well-drained soils. Note that such soils tend to have greater slopes and soil K values than the more poorly drained soils, which makes them more susceptible to erosion. Adoption of conservation tillage because of the economic benefits of greater yield can, therefore, help to reduce soil erosion. All forms of tillage other than moldboard plowing (Table 9) increase the concentration of crop residue on the surface or buried near the surface, but may not have 30 percent of the surface covered with crop residue, depending on the amount of secondary tillage.

Soil drainage criteria and soil taxonomy are among the most important soil factors used in the Corn Belt to guide the adoption of conservation tillage in an area the size of a tillage management region (*3*). These guides may indeed be a partial response to the dilemma posed above regarding aggregation of land resource areas up to a tillage management region. Cosper (*26*) used soil taxonomy (soil moisture in the control section; aquic, aridic, udic, ustic, and xeric as designated in the subgroup category) to predict where the three broadly described tillage forms could be used for corn production in the Corn Belt. These forms were all forms of tillage, including no-till; reduced and conventional tillage, excluding no-till; and conventional tillage only.

Several states in the Corn Belt have elaborate and current guides to recommend tillage systems related to soil properties. Galloway and associates (*43*) suggested nine forms of tillage-planting for corn and soybean on 23 groups of soil series in Indiana. They considered seven limitations: ponded water, groundwater table, perched water table, stream overflow, wind erosion, water erosion, and slope steepness in the upland position. In these limitations they used soil clay content to indicate internal drainage and timeliness of field drying and organic matter to determine stability of surface soil structure and susceptibility to soil erosion (*42*).

Iowa (*78*) uses six soil-related factors (soil wetness and low soil temperature, potential water/wind erosion, potential for flooding, topsoil thickness, and soil tilth expressed as a ratio of organic matter to clay content) to rate performance of select tillage systems. The tillage systems include full-width, strip-till, and no-till, which were selected on the basis of their use to address the soil factors on a given soil series. This system has been expanded (G. A. Miller, personal communication) to include 10 specific tillage-planting systems.

Three other states (Minnesota, Michigan, and Ohio) in the Corn Belt have recommended various forms of tillage based upon internal and surface-soil drainage (*3*). Minnesota selected a tillage system to manage problems of soil crusting, low soil temperature, timeliness of tillage and planting, surface roughness, soil erosion, stored water potential, and soil compaction in each of six broad soil groups. Ohio considered five soil groups evaluated for drainage and surface clay content (as a measure of surface crusting and retarded infiltration). Michigan considered slope and drainage class primarily for success of no-till.

Conservation Tillage in Double-Cropping

The double-cropping system is clearly facilitated by conservation tillage. Conservation tillage usage, principally NT for the summer crop and MT for the winter crop, was noted in the Southern Corn Belt, Piedmont, Eastern Uplands, and Coastal Plains tillage management regions. Double-cropping must be used in arable agriculture to control soil erosion on Ultisols and Alfisols of the southeastern United States (*104, 109*), yet the number of hectares of double-cropping varies widely due to commodity prices and other external factors (*75*). Based upon the moisture and temperature regime of the taxonomic control section, these are mostly udic-thermic. The only other form of arable management suitable for erosion-susceptible Ultisols is a sod-based crop rotation (*18, 104*). Even in this system, special techniques, such as contour balk, mulch tillage, lister tillage, and wheel-track planting were required to control soil erosion (*104, 109*). However, adoption still did not proceed until double-cropping technology was developed during the 1960s (*7*).

Numerous technological innovations facilitated crop production with crop residues on the surface in double-cropping. Among these are the fluted coulter and in-row subsoiler interfaced with planters and a flexible herbicide technology, including application techniques and modes of herbicide action (*104, 109*). When it was found that sod stands of cool-season crops were depleted after 3 to 5 years of double-cropping with a summer crop (*104, 109*), research was initiated recently to produce cool-season and cereal grain legumes as cover crops for mulch purposes (*63, 88*). Commodity market conditions and agricultural chemicals of lower cost and improved efficacy have encouraged small grain production. However, the latter system of small grain harvest with a summer crop (soybean or grain sorghum) is sustainable because of significant nitrogen retention in crop residue (as great as 12 megagrams per hectare per year) and the projected low potential for soil erosion (*62, 64*). Throughout the development of double-cropping, a mild cool season and very short periods without surface residues and vegetative cover have repeatedly posed disease and insect problems that could change the rotation. For instance, take-all (*Gaeumannomyces graminis* var. *tritici*) in winter wheat is suppressed by grain sorghum rather than soybean as a summer crop (*92*).

Double-cropping technology is available for application, but the hectares of soybean-wheat double-cropping varies greatly because of market conditions (especially that of wheat) and rainfall deficiencies. There has been a 20 to 30 percent decline in double-cropping soybean since the 1982 high (*75*). If wheat prices remain low, the potential for soybean (double-cropping) conversion to soybean full season would indeed be detrimental to soil conservation and renewed production of Ultisols and Alfisols that had been severely eroded before the adoption of double-cropping and conservation tillage. The drouth in 1988 and 1989 increased the use of no-till at the expense of strip-till because the in-row subsoiler was used less.

Irrespective of plant disease and marketing problems associated with double-cropping on Ultisols and Alfisols, rapidly advancing technology for residue management (*109*) and long-term influences of conservation tillage (*63*) continue to provide potential for double-cropping and associated conservation. A long-term continuum of multiple crops in a conservation tillage mode is necessary. Crop rotation is, however, essential for sustainability.

New Tillage Systems in the Southern Plains

The CTIC national survey of conservation tillage for 1985 indicates that 30 percent of the cropland in the Southern Great Plains tillage management region was in some form of conservation tillage. Of the total, 4 percent was in NT, 1 percent in RiT, 55 percent in MT, and 40 percent in ReT. The amount in ST was less than 0.5 percent. In contrast, the survey indicates that some form of CT was used on about 17 percent of the cropland in Texas in 1986, including about 7 percent NT, 2 percent RiT, less than 1 percent ST, 85 percent MT, and 6 percent ReT (*48*). The totals do not include cropland that is in permanent pasture, in fallow for the entire year, or in conservation use (set-aside or diverted areas). Including these areas reduces the total area of CT for Texas to about 14 percent.

The discrepancy between CT use in the Southern Great Plains tillage management region (30 percent) and in Texas (17 percent) is related to several factors, foremost of which is that the Southern Great Plains tillage management region is a composite of several well-defined tillage management regions. Although a major portion of Texas is included in the Southern Great Plains as defined by Jones and Johnson (*54*), major areas of cropland in New Mexico, Colorado, Kansas, and Oklahoma also are included in the Southern Great Plains tillage management region. Much of this cropland is used for small grain (primarily winter wheat) or grain sorghum production, usually in some combination with fallow at least one year in three, for which MT or ReT methods are quite common. The most widely used implements are stubble mulch tillage (V-sweep or blade) and heavy-duty cultivators (chisel type) in the drier, nonirrigated portions of these states. In Texas CT is used for about 28 to 13 percent of wheat and grain sorghum, respectively. A similar use of CT for wheat in other states is expected in the Southern Great Plains.

Under irrigated conditions and in more humid portions of the Southern Great Plains tillage management region where annual cropping of small grains is common (*27*), offset disking and chiseling are widely practiced. Also, moldboard plowing is common in more humid areas, especially in parts of Oklahoma, Kansas, and Texas where high residue amounts result in weed control and seeding problems. Bull (*12*) showed that there was much use of conventional tillage, with or without moldboard plowing, but practically no use of mulch till.

A second major reason for the discrepancy in the use of CT between the Southern Great Plains tillage management region and Texas is that the Texas survey includes several major subregions that are not part of the Southern Great Plains, as defined by Jones and Johnson (*54*), but use little CT. These are the Upper Coast Prairie, Coastal Bend, and Lower Valley major land resource areas, where CT tillage use is 10, 16, and 9 percent, respectively. Low use of CT in these subregions, however, is acceptable because the topography is relatively flat and not susceptible to erosion. A major impact on the Texas total is the little use of CT for cotton production. Dryland cotton in Texas is grown in nearly monoculture (*27*), and CT is used only on about 2 percent of that area. Most of the cotton in Texas is grown in the Southern High Plains and Southern Rolling Plains major land resource areas, where total use of CT is about 8 to 12 percent, respectively, which is considerably less than the Texas total and much lower than where grain crops are primarily grown.

CT is infrequently used for cotton because cotton residues (stalks) remaining after lint harvest are sufficiently sparse that a 30 percent cover cannot be maintained to control erosion; most cotton is produced on soils that are highly erodible and require soil roughening or ridge tillage to reduce the potential for erosion; and the relatively high economic returns for cotton as compared with other crops encourages cotton monoculture.

Because of the low residue amounts produced by cotton and the highly erodible sandy soils on which much of the cotton is grown, tillage for cotton frequently is lister tillage that covers virtually all remaining residues and results in a ridge and furrow surface configuration that is important for wind erosion control. Often, lister tillage is preceded by moldboard plowing at a depth of 20 to 25 centimeters to bring less erodible, cloddy soil materials to the surface or by chiseling, which roughens the surface sufficiently to help control erosion. Consistence of erosion control is somewhat haphazard because the effectiveness of any tillage practice for controlling wind erosion is highly dependent on the soil water content at the time of tillage and the particular soil.

Although use of CT in the Southern Great Plains tillage management region is lower than in most other tillage management regions, its use is expanding in many subregions of the Southern Great Plains tillage management region due to continuing advances in technology (*48*). For the drier grain-producing subregions, where use of CT is already higher than for other subregions, a 3-year rotation of a winter crop of small grain (usually winter wheat) and a summer crop of grain sorghum, corn, or millet (*Panicum miliaceum* L.) has potential for increased use of conservation tillage (*84, 107, 108*). This system results in two crops in 3 years, with about 11 months of fallow between each crop. A conservation tillage, especially no-till during the fallow after wheat harvest, has increased soil water storage, as much as 75 millimeters in many cases, and increased grain yields, as much as 1.0 megagram per hectare in some cases, as compared with a clean tillage system. Besides greater water storage and grain yields, the above rotations also improved weed control and usually provided adequate surface residues for controlling erosion, especially where a no-till system was used.

Conservation tillage (no-till with herbicides or tillage-herbicide combinations) also has increased water storage and grain yields in a wheat-fallow system (*53, 80, 84*). The increased water storage and yields were attributed to greater surface residue retention with conservation tillage as compared with clean tillage.

In more humid portions of the Southern Great Plains tillage management region, where monoculture of winter wheat is common and where large amounts of crop residue on the surface are problematic for weed control and planting, moldboard plowing was used exclusively until recently. Now, offset disking and chiseling are common, with occasional moldboard plowing (*12*) or even burning to help control grassy weeds, such as cheatgrass (*Bromus secalinus* L.). Although reduced use of moldboard plowing enhances the potential for resource conservation through greater crop residue retention near the surface, frequent secondary tillage leaves the surface virtually devoid of residues at wheat planting time (note the estimated residue cover of 2 to 11 percent for conventional tillage in Oklahoma) (*12*).

For cotton-producing subregions, where there is little use of CT, recent studies have shown that some conservation tillage systems can improve cotton lint yields and net returns (*55, 71*). If associated economic returns are favorable, producers will undoubtedly start using such conservation tillage systems because they may be required to adopt some form of conservation tillage if they are to comply with provisions of the 1985 Food Security Act (D. W. Fryrear, 1989, personal communication). To achieve even greater adoption of conservation tillage, especially for cotton and wheat monoculture, continued research is needed to develop more widely effective and acceptable technology.

Summary and Conclusions

Major trends in adoption of conservation tillage-planting systems in the conterminous United States in 1985 were readily described by the statistics gathered by the Conservation Technology Information Center, especially when supported by published scientific literature about the performance and character of the various tillage systems. However, the survey has missed some very significant changes in adoption of tillage-planting systems.

Eleven tillage management regions were suggested for aggregation of survey information taken nationally at the county level. These were intended to characterize soil, climate, adapted crop, and cultural factors controlling the adoption of specific conservation tillage-planting systems. Five conservation tillage-planting systems (CT), each with 30 percent of the surface covered with crop residue after planting, were considered, including mulch tillage (MT), no-till (NT), reduced till (ReT), ridge-till (RiT), and strip-till (ST). Major crops (and categories of crops) planted were small grain, soybean, grain sorghum, corn, cotton, "vegetable and truck crops," "forage crops," and "other crops." Soybean, corn, and grain sorghum were considered as full season or the summer crop in a double-cropping rotation.

CT ranged from a low of 11 percent in the Associated Delta to a high for 47 percent for nonforage crops planted in the Piedmont tillage management region. Adoption of conservation tillage greater than 38 percent was noted in the Southern Corn Belt, Eastern Uplands, Central Great Plains, and Piedmont tillage management regions. The adoption of partial-width forms of conservation tillage (ST, NT, and RiT) were small compared with adoption of full-width forms (MT and ReT). NT ranged from about 1 to 4 percent of planted hectares in the Northern Great Plains and Northern Corn Belt and from 7 to 15 percent in the Southern Corn Belt, Eastern Uplands, Coastal Plains, and Piedmont tillage management regions. ST showed some adoption in the Coastal Plains and Central Great Plains, while RiT is starting to be adopted in the Northern/Southern Corn Belt and Central Great Plains tillage management regions. MT ranged from a high of 30 percent for nonforage crops planted in the Northern/Southern Corn Belt to a low of 4 to 7 percent in the Associated Delta and Northeast tillage management regions. ReT (a full-width system in which the primary tillage and the planting operations are simultaneous to eliminate secondary tillage) was used most frequently in the Central Great Plains, Southern Great Plains, Piedmont, and Eastern Uplands tillage management regions.

Adoption of CT is highest in those tillage management regions in which numerous crops are adapted (Central Great Plains, Piedmont, Eastern Uplands), an association expected if crop rotation were a major factor in the application of conservation tillage. In contrast, adoption of CT in the Pacific Northwest and Northern Great Plains was lower because so few crops are adapted to these tillage management regions and a large fraction of the cropland is summer-fallowed.

Interpretation of the adoption of various forms of CT for each of the major crops within a tillage management region was difficult without extreme reliance on published literature to clarify the importance of such factors as tillage rotation, crop rotation, and soil property, factors not elucidated in the survey. However, numerous relationships were clearly demonstrated, one of which is the use of NT for summer crops and MT for winter crops in the double-cropping management.

Adoption of conservation tillage systems is driven by economics and risk, mainly of the short-term range, and by a lack of technology that may be integrated into a productive system. Conservation tillage systems that do not use the moldboard plow generally have lower cost inputs and, when there is no risk of lower crop yields, adoption of conservation tillage is clearly indicated. In the two Corn Belt tillage management regions and adjacent tillage management regions, all forms of conservation tillage depressed soybean and corn yields on fine-textured and poorly drained soils, whereas yields of these crops were enhanced by conservation tillage on imperfectly drained and well-drained soils.

Conservation tillage technology requires a compatible combination of cultural pest controls and equipment for precise and timely application of fertilizers, for handling crop residues without deep incorporation and fragmentation, and for precise placement of seeds in a seedbed with trashy and unconventional tilth. Although gains in soil and water conservation (control of wind/water erosion, enhanced water relations through improved infiltration, and reduced evaporation) are well documented and recognized by the public, they are secondary factors except where crop yields are improved.

Unfortunately, most economic and risk analyses have been derived from experimentation in which the same tillage system has been repeated continuously even for those analyses involving crop rotation (e.g., corn and soybean), whereas most farm fields use tillage rotation.

Crop residue cover on the soil surface, crop residue buried in the upper 10 centimeters, and surface roughness (random and oriented) are all important factors for controlling water/wind erosion and improving soil water relations. All three factors can be quantified to predict soil erosion control by conservation tillage; the relation to soil water relations is semiqualitative.

Complete reliance on surface cover (with crop residue) as a characteristic to ascertain a conservation tillage-planting system in the survey of adoption has overlooked some important trends in the evolution of full-width conservation tillage systems. Several surveys in the two Corn Belt and the Central Great Plains tillage management regions and a recent national survey by the Economic Research Service show that chisel plows and disks are replacing moldboard plows for primary tillage, yet the trend has not shown up in the CTIC survey because there has been sufficient secondary tillage to reduce the surface cover to less than 30 percent after planting. However, this change has maintained 80 to 90 percent of the crop residue in the 0- to 10-centimeter depth instead of the 40 percent or less at that depth with moldboard plowing.

In some parts of the United States, random and oriented soil roughness can be used to reduce soil erosion (wind/water) and to enhance water conservation. Random roughness is sensitive to soil texture, primary tillage tool, and the amount of secondary tillage. There are guides for semiquantitative field estimation and estimation of effectiveness in the control of soil erosion and soil water relations.

To obtain a better knowledge of the adoption, benefits, and technological needs of conservation tillage-planting systems, the CTIC surveys must include the following:

1. Survey results are needed at the county level and must be aggregated consistent with natural features that control adoption. The suggested tillage management regions were somewhat successful, but a better delineation is needed, based upon predicted plant response to soil factors and stochastic climatic inputs. Soil factors could be made up of some assortment of factors now used in tillage guides in the Corn Belt. Other specific soil factors were discussed previously.

2. Information at the county level must indicate crop rotations and tillage rotations in relation to the various conservation tillage-planting systems.

3. Crop classifications of "forage," "vegetable and truck," and "other" do not aggregate crop species with a common advantage/disadvantage for conservation tillage-planting systems. Root crops generally present a problem of 30 percent residue cover in the subsequent crop unless a cover crop is used. Cover crops of various species should be grouped, provided they are not also the summer crop, as with full-season winter wheat. Some crops, such as sunflower and pea, should be identified because they are suitable in conservation tillage systems.

4. The partial-width tillage systems of NT, ST, and RiT are adequately described, but the full-width systems require more clarification to interpret the presence of shallow residue burial, length of time residues are on the surface, and management of surface roughness. All three aspects of full-width tillage management are elements of soil erosion control and improved soil water relations. There is a need to know if (a) a moldboard plow or a disk/chisel plow was used for primary tillage, (b) the primary tillage was done in fall or spring, (c) a quantity of random/oriented roughness was maintained, and (d) primary tillage was delayed until coincident with planting. Evaluating only surface residue cover after planting implies wrongly that the only serious soil erosion occurs after planting and the only control is surface residue cover.

A new restraint on the development and use of conservation tillage systems will be control of groundwater contamination resulting from the use of agricultural chemicals. It is too early to predict the influence on tillage systems, but one can expect more use of combined cultural and chemical methods for pest control. For instance, plant disease control may influence whether the residue remains as a surface cover, is buried just below the surface, or must be deeply incorporated, as with a moldboard plow. This decision would depend upon whether the inoculum is in the shoot residue, root residue, or both. A large carryover of crop residue incorporated at a shallow depth may have a significant impact on herbicide activity and persistence as compared with the other options of surface residue or deeply incorporated residue. There is more shallow-incorporated crop residue, as with the switch to disks and chisel plows. Adjustments in fertilizer application as related to residue positioning will be needed to sustain adequate nutrition yet control movement of these constituents from the root zone via surface runoff or leaching.

REFERENCES CITED

1. Allen, R. R., and C. R. Fenster. 1986. *Stubble-mulch equipment for soil and water conservation in the Great Plains.* Journal of Soil and Water Conservation 41: 11-16.
2. Allmaras, R. R., P. M. Burford, J. L. Pikul, Jr., and D. E. Wilkins. 1988. *Tillage tool influences on incorporated wheat residues.* In 11th International Conference Proceedings, Volume 1. International Soil Tillage Research Organization, Haren, The Netherlands. pp. 445-450.
3. Allmaras, R. R., P. W. Unger, and D. E. Wilkins. 1985. *Conservation tillage systems and soil productivity.* In R. F. Follett and B. A. Stewart [editors] *Soil Erosion and Crop Productivity.* American Society of Agronomy, Madison, Wisconsin. pp. 357-413.

4. Allmaras, R. R., and R. H. Dowdy. 1985. *Conservation tillage systems and their adoption in the United States.* Soil and Tillage Research 5: 197-222.
5. American Society of Agricultural Engineers. 1988. *Standards 1988.* American Society of Agricultural Engineers, St. Joseph, Michigan. 629 pp.
6. Bauer, A., A. B. Frank, and A. L. Black. 1984. *Estimation of spring wheat leaf growth rates and anthesis from air temperature.* Agronomy Journal 76: 829-835.
7. Beale, O. W., and G. W. Langdale. 1967. *Tillage and residue management practices for soybean production in a soybean-small grain rotation.* Agronomy Journal 59: 31-33.
8. Black, A. L. 1973. *Soil property changes associated with crop residue management in a wheat-fallow rotation.* Soil Science Society of America Proceedings 29: 465-468.
9. Brown, H. J., R. M. Cruse, and T. S. Colvin. 1989. *Tillage system effects on crop growth and production costs for a corn-soybean rotation.* Journal of Production Agriculture 2: 273-279.
10. Brown, L. C., G. R. Foster, and D. B. Beasley. 1989. *Rill erosion as affected by incorporated residue and seasonal consolidation.* Transactions, American Society of Agricultural Engineers 32: 1967-1968.
11. Bull, L. 1989. *Residue and tillage systems in 1987 corn production.* In *Agricultural Resources: Inputs, Situation, and Outlook.* AR-13. Economic Research Service, Washington, D.C. pp. 36-41.
12. Bull, L. 1989. *Tillage practices and residue levels.* In *Agricultural Resources: Inputs, Situation, and Outlook.* AR-15. Economic Research Service, U.S. Department of Agriculture, Washington, D.C. pp. 6-14.
13. Bureau of the Census. 1982. *1978 Census of Agriculture, Volume 5, Special reports, Part 1, graphic summary.* U.S. Department of Commerce, Washington, D.C. 177 pp.
14. Burnside, O. C., G. A. Wicks, and D. R. Carlson. 1980. *Control of weeds in an oat (Avena sativa)-soybean (Glycine max) ecofarming rotation.* Weed Science 28: 46-50.
15. Burwell, R. E., and W. E. Larson. 1969. *Infiltratrion as influenced by tillage-induced random roughness and pore space.* Soil Science Society of America Proceedings 33: 449-452.
16. Campbell, R. B., D. L. Karlen, and R. E. Sojka. 1984. *Conservation tillage for maize in the U.S. Southeastern Coastal Plains.* Soil and Tillage Research 4: 511-529.
17. Campbell, R. B., R. E. Sojka, and D. L. Karlen. 1984. *Conservation tillage for soybeans in the U.S. Southeastern Coastal Plains.* Soil and Tillage Research 4: 531-541.
18. Carreker, J. R., S. R. Wilkinson, A. P. Barnett, and J. E. Box. 1977. *Soil and water management systems for sloping land.* ARS-S-160. Agricultural Research Service U.S. Department of Agriculture, Washington, D.C. 76 pp.
19. Cassel, D. K. 1983. *Effects of soil characteristics and tillage practices on water storage and its availability to plant roots.* In C. D. Raper, Jr., and P. J. Kramer [editors] *Crop Reactions to Water and Temperature Stresses in Humid, Temperate Climates.* Westview Press, Boulder, Colorado. pp. 167-186.
20. Colvin, T. S., J. M. Laflen, and D. C. Erbach. 1981. *A review of residue reduction by individual implements.* In J. C. Siemens [editor] *Crop Production with Conservation in the 80s.* ASAE Publication 7-81. American Society of Agricultural Engineers, St. Joseph, Michigan. pp. 102-110.
21. Conservation Technology Information Center. 1989. *1988 national survey: Conservation tillage practices.* West Lafayette, Indiana. 144 pp.
22. Conservation Tillage Information Center. 1984. *Conservation tillage: Local action and views: A survey of America's Conservation Districts.* West Lafayette, Indiana. 19 pp.
23. Conservation Tillage Information Center. 1986. *1985 national survey: Conservation tillage practices.* West Lafayette, Indiana. 136 pp.
24. Cook, R. J. 1984. *Root health: Importance and relationship to farming practices.* In D. F. Bezdicek et al. [editors] *Organic Farming: Current Technology and Its Role in a Sustainable Agriculture.* Special Publication No. 46. American Society of Agronomy, Madison, Wisconsin. pp. 111-127.
25. Cook, R. J., M. G. Boosalis, and B. Doupnik. 1978. *Influence of crop residues on plant diseases.* In W. R. Oschwald et al. [editors] *Crop Residue Management Systems.* Special Publication No. 31. American Society of Agronomy, Madison Wisconsin. pp. 147-163.
26. Cosper, H. R. 1983. *Soil suitability for conservation tillage.* Journal of Soil and Water Conservation 38: 152-155.
27. Daberkow, S., and M. Gill. 1989. *Common crop rotations among major field crops.* In *Agricultural Resources: Inputs, Situation, and Outlook.* AR-15. Economic Research Service, U.S. Department of Agriculture Washington, D.C. pp. 34-40.
28. Deibert, E. J. 1989. *Reduced tillage systems influences on yield of sunflower hybrids.* Agronomy Journal 81: 274-279.
29. Dickey, E. C., P. J. Jasa, B. J. Dolesh, L. A. Brown, and S. K. Rockwell. 1987. *Conservation tillage: Perceived and actual use.* Journal of Soil and Water Conservation 42: 431-434.
30. Doster, D. H., D. R. Griffith, J. V. Mannering, and S. D. Parsons. 1983. *Economic returns from alternative corn and soybean tillage systems in Indiana.* Journal of Soil and Water Conservation 38: 504-508.
31. Dregne, H. E., and W. O. Willis, editors. 1983. *Dryland agriculture.* Agronomy Monograph 23. American Society of Agronomy, Madison, Wisconsin. 622 pp.
32. Duffy, M., and M. Hanthorn. 1984. *Returns to corn and soybean tillage practices.* Agricultural Economics Report No. 508. Economic Research Service, U.S. Department of Agriculture, Washington, D.C. 14 pp.
33. Elkins, C. B., D. L. Thurlow, and J. G. Hendrick. 1983. *Conservation tillage for long-term amelioration of plow pan soils.* Journal of Soil and Water Conservation 38: 305-307.
34. Epplin, F. M., and T. F. Tice. 1986. *Influence of crop and farm size on adoption of conservation tillage.* Journal of Soil and Water Conservation 41: 424-427.
35. Erbach, D. C. 1982. *Tillage for continuous corn and corn-soybean rotation.* Transactions, American Society of Agricultural Engineers 25: 906-911, 918.
36. Erbach, D. C., J. E. Morrison, and D. E. Wilkins. 1983. *Equipment modification and innovation for conservation tillage.* Journal of Soil and Water Conservation 38: 182-185.
37. Fletcher, J. J., 1985. *Corn Belt survey results: Impact of changing tillage practices on the fertilizer industry.* Bulletin Y-187. National Fertilizer Development Center.

Tennessee Valley Authority, Muscle Shoals, Alabama. pp. 77-87.

38. Forcella, F., and M. J. Lindstrom. 1988. *Movement and germination of weed seeds in ridge-till crop production systems.* Weed Science 36: 56-59.
39. Forcella, F., and M. J. Lindstrom. 1988. *Weed seed and populations in ridge and conventional tillage.* Weed Science 36: 500-503.
40. Frye, W. W., R. L. Blevins, M. S. Smith, S. J. Corak, and J. J. Varco. 1988. Role of annual legume cover crops in efficient use of water and nitrogen. In W. L. Hargrove [editor] *Cropping Strategies for Efficient Use of Water and Nitrogen.* Special Publication No. 51. American Society of Agronomy, Madison, Wisconsin. pp. 129-154.
41. Fryrear, D. W. 1985. *Soil cover and wind erosion.* Transactions, American Society of Agricultural Engineers 28: 781-784.
42. Galloway, H. M., D. R. Griffith, and J. V. Mannering. 1981. *Adaptability of various tillage-planting sysems to Indiana soils.* Cooperative Extension Service Bulletin AY-210. Purdue Universitty, West Lafayette, Indiana. 15 pp.
43. Galloway, H. M., D. R. Griffith, and J. V. Mannering. 1986. *Adaptability of various tillage-planting sysems to Indiana soils.* Cooperative Extension Service Bulletin AY210, Revised. Purdue University, West Lafayette, Indiana. 15 pp.
44. Gregory, J. M. 1982. *Soil cover prediction with various amounts and types of crop residues.* Transactions, American Society of Agricultural Engineers 25: 1,333-1,337.
45. Griffith, D. R., J. V. Mannering, H. M. Galloway, S. E. Parsons, and C. B. Richey. 1973. *Effect of eight tillage planting systems on soil temperature, percent stand, plant growth, and yield of corn on five Indiana soils.* Agronomy Journal 65: 321-326.
46. Hargrove, W. L., and W. S. Hardcastle. 1984. *Conservation tillage practices for winter wheat production in the Appalachian Piedmont.* Journal of Soil and Water Conservation 39: 324-326.
47. Harman, W. L., and A. F. Wiese. 1985. *Economic appraisal of reduced-tillage practices.* In A. F. Wiese [editor] *Weed Control in Limited-Tillage Systems.* Weed Science Society of America, Champaign, Illinois. pp. 282-297.
48. Harris, B. L., E.C.A. Runge, and G. K. Westmoreland. 1987. *Conservation tillage systems in Texas.* In T. J. Gerik and B. L. Harris [editors] *Conservation Tillage: Today and Tomorrow.* Miscellaneous Publication MP-1636. Texas Agricultural Experiment Station, College Station. pp. 1-7.
49. Jeffers, D. L. 1984. *New ways to double-crop soybeans.* Proceedings, 13th Soybean Research Conference, American Seed Trade Association 123: 83-90.
50. Johnson, R. R. 1987. *Crop management.* In J. R. Wilcox [editor] *Soybeans: Improvement, Production, and Uses* (2nd edition). Agronomy Monograph 16. American Society of Agronomy, Madison Wisconsin. pp. 355-390.
51. Johnson, R. R. 1988. *Soil engaging-tool effects on surface residue and roughness with chisel-type implements.* Soil Science Society of America Journal 52: 237-243.
52. Jolly, R. W., W. M. Edwards, and D. C. Erbach. 1983. *Economics of conservation tillage in Iowa.* Journal of Soil and Water Conservation 38: 291-294.
53. Jones, O. R., A. F. Wiese, and G. L. Johnson. 1989. *Tillage and sequence effects on dryland crop production.* In P. W. Unger, W. R. Jordan, T. V. Sneed, and R. W. Jensen [editors] *Proceedings, International Conference on Dryland Farming.* Texas A&M University, College Station.
54. Jones, O. R., and W. C. Johnson. 1983. *Cropping practices: Southern Great Plains.* In H. E. Dregne and W. O. Willis [editors] *Dryland Agriculture.* Agronomy Monograph 23. American Society of Agronomy, Madison, Wisconsin. pp. 365-384.
55. Keeling, J. W., C. W. Wendt, J. R. Gannaway, A. B. Onken, W. M. Lyle, R. J. Lascano, and J. R. Abernathy. 1988. *Conservation tillage cropping systems for the Texas Southern High Plains.* In K. H. Remy [editor] *Proceedings, 1988 Southern Conservation Tillage Conference.* Special Bulletin 88-1. Mississippi Agriculture and Forest Experiment Station, Mississippi State. pp. 19-21.
56. Keeling, W., E. Segarra, and J. R. Abernathy. 1989. *Evaluation of conservation tillage cropping systems for cotton on the Texas Southern High Plains.* Journal of Production Agriculture 2: 269-273.
57. Kohler, M. A., T. J. Nordenson, and D. R. Banker. 1959. *Evaporation maps for the United States.* Technical Paper No. 37. U.S. Department of Commerce, Washington, D.C.
58. Klemme, R. M. 1985. *A stochastic dominance comparison of reduced tillage systems in corn and soybean production under risk.* American Journal of Agriculture Economics 67: 550-557.
59. Ladewig, H., and R. Gariby. 1983. *Reasons why Ohio farmers decide for or against conservation tillage.* Journal of Soil and Water Conservation 38: 487-488.
60. Laflen, J. M., G. R. Foster, and C. A. Onstad. 1985. *Simulation of individual storm soil loss for modeling the impact of soil erosion on soil productivity.* In S. A. El-Swaify et al. [editors] *Soil Erosion and Conservation.* Soil Conservation Society of America, Ankeny, Iowa.
61. Langdale, G. W. H. F. Perkins, A. P. Burnett, J. C. Reardon, and R. L. Wilson, Jr. 1983. *Soil and nutrient runoff losses with in-row chisel planted soybeans.* Journal of Soil and Water Conservation 38: 297-301.
62. Langdale, G. W., and R. L. Wilson, Jr. 1987. *Intensive croping sequences to sustain conservation tillage for erosion control.* Journal of Soil and Water Conservation 42: 352-355.
63. Langdale, G. W., R. W. Wilson, Jr., and R. R. Bruce. 1990. *Cropping frequencies to sustain long-term conservation tillage systems.* Soil Science Society of America Journal 54: 193-198.
64. Langdale, G. W., W. L. Hargrove, and J. Giddens. 1984. *Residue management in double crop conservation tillage systems.* Agronomy Journal 76: 689-694.
65. Larson, W. E. 1964. *Soil parameters for evaluating tillage needs and operations.* Soil Science Society of America Proceedings 28: 119-122.
66. Larson, W. E., and J. J. Hanway. 1977. *Corn production.* In G. F. Sprague [editor] *Corn and Corn Improvement.* Agronomy Monograph 18. American Society of Agronomy, Madison, Wisconsin. pp. 625-669.
67. Larson W. E., and O. W. Beale. 1961. *Using crop residues on soil of the humid area.* Farmers Bulletin No. 2155. U.S. Department of Agriculture, Washington, D.C. 14 pp.
68. Lersten, N. R., and J. B. Carlson. 1987. *Vegetative morphology.* In J. R. Wilcox [editor] *Soybeans: Improvement, Production, and Uses (second edition).* Agronomy

Monograph 16. American Society of Agronomy, Madison, Wisconsin. pp. 49-94.

69. Lewis, W. M., and J. A. Phillips. 1976. *Double cropping in eastern United States.* In R. I. Papendick et al. [editors] *Multiple Cropping.* Special Publication No. 27. American Society of Agronomy, Madison, Wisconsin. pp. 41-50.
70. Lowdermilk, W. C. 1953. *Conquest of the land through 7000 years.* Infomation Bulletin No. 99. U.S. Department of Agriculture, Washington, D.C. 30 pp.
71. Lyle, W. M., and J. P. Bordovsky. 1987. *Integrating irrigation and conservation tillage technology.* In T J. Gerik and B. L. Harris [editors] *Conservation Tillage: Today and Tomorrow.* Miscellaneous Publication MP-1636. Texas Agricultural Experiment Station, College Station pp. 67-71.
72. Mannering, J. V., and C. R. Fenster. 1983. *What is conservation tillage?* Journal of Soil and Water Conservation 38: 141-143.
73. Mannering, J. V., and D. R. Griffith. 1981. *Value of crop rotation under various tillage systems.* Cooperative Extension Bulletin AY-230. Purdue University, Lafayette, Indiana 5 pp.
74. Mannering, J. V., D. R. Griffith, and K. D. Johnson. 1985. *Winter cover crops—their value and management.* Cooperative Extension Bulletin AY-247. Purdue University, Lafayette, Indiana. 3 pp.
75. Marra, M. C., and G. A. Carlson. 1986. *Double-cropping wheat and soybeans in the Southeast: Input use and pattern of adoption.* Agricultural Economic Report 552. Economic Research Service, U.S. Department of Agriculture, Washington, D.C. 18 pp.
76. Martin, J. H., W. H. Leonard, and D. L. Stamp. 1976. *Principles of field crop production (third edition).* Macmillan, New York, New York. 1,118 pp.
77. McCalla, T. M., and T. J. Army. 1961. *Stubble mulch farming.* Advances in Agronomy 13: 125-196.
78. Miller, G. A., T. E. Fenton, and R. M. Cruse. 1982. *Full width tillage systems: Soil adaptability.* In *Farm Agricultural Resources Management.* Iowa State University, Ames.
79. Moody, J. E., G. M. Shear, and J. N. Jones, Jr. 1961. *Growing corn without tillage.* Soil Science Society of America Proceedings 25: 516-517.
80. Morishita, D. W., A. J. Schlegel, C. A. Norwood, and R. E. Gwin. 1989. *Comparison of herbicides and tillage in a long-term wheat-fallow rotation.* In P. W. Unger, W. R. Jordan, T. V. Sneed, and R. W. Jensen [editors] *Proceedings, International Conference on Dryland Farming.* Texas A&M University, College Station.
81. Morrison, J. E., Jr., S. H. Parker, C. A. Jones, R. R. Allen, D. E. Wilkins, and others. 1989. *Expert system for selecting conservation tillage planting machines: PLANTING.* Transactions, American Society of Agricultural Engineers 32: 397-401.
82. Moschler, W. W., D. C. Martens, C. I. Rich, and G. M. Shear. 1973. *Comparative lime effects on continuous no-tillage and conventionally tilled corn.* Agronomy Journal 65: 781-783.
83. Musgrave, R. B., P. J. Zwerman and S. R. Aldrich. 1955. *Plow planting of corn.* Agriculture Engineering 36: 593-594.
84. Norwood, C. A., A. J. Schlegel, D. W. Morishita, and R. E. Gwin. 1989. *Reduced tillage cropping systems in southwest Kansas.* In P. W. Unger, W. R. Jordan, T. V. Sneed, and R. W. Jensen [editors] *Proceedings, International Conference on Dryland Farming.* Texas A&M University, College Station.
85. Office of Technology Assessment. 1982. *Impacts of technology on U.S. cropland and rangeland productivity.* U.S. Congress, Washington, D.C. 266 pp.
86. Phillips, R. E., R. L. Blevins, G. W. Thomas, W. W. Frye, and S. H. Phillips. 1980. *No-tillage agriculture.* Science 208: 1,108-1,113.
87. Pierce, F. J., R. H. Dowdy, W. E. Larson, and W.A.P. Graham. 1984. *Soil productivity in the Corn Belt: An assessment of erosion's long-term effects.* Journal of Soil and Water Conservation 39: 131-136.
88. Power, J. F., editor. 1987. *The role of legumes in conservation tillage systems.* Soil and Water Conservation Society, Ankeny, Iowa.
89. Ramig, R. E., and D. E. Smika. 1954. *Fallow-wheat-sorghum: An excellent rotation for dryland in central Nebraska.* Bulletin No. 483. Nebraska Agricultural Experiment Station, Lincoln. 11 pp.
90. Randall, G. W., J. B. Swan, and D. T. Walters. 1984. *Report on long-term conservation tillage studies at Waseca, Minnesota.* Agricultural Experiment Station Miscellaneous Publication No. 2. Department of Soil Science, University of Minnesota, St. Paul. pp. 113-117.
91. Reicosky, D. C., D. K. Cassell, R. L. Blevins, W. R. Gill, and G. C. Naderman. 1977. *Conservation tillage in the Southeast.* Journal of Soil and Water Conservation 32: 13-19.
92. Rothrock, C. S., and G. W. Langdale. 1989. *Influence of nonhost summer crops on take-all in double-cropped winter wheat.* Plant Disease 73: 130-132.
93. Sanford, J. E., D. L. Myhre, and N. C. Merwine. 1973. *Double cropping systems involving no-tillage and conventional tillage.* Agronomy Journal 65: 978-982.
94. Sanford, J. O., W. P. Williams, J. E. Hairston, and L. L. Reinschmiedt. 1988. *Double cropping insect and disease resistant corn with wheat.* Journal of Production Agriculture 1: 60-63.
95. Schertz, D. L. 1988. *Conservation tillage: An analysis of acreage projections in the United States.* Journal of Soil and Water Conservation 43: 256-258.
96. Shaw, R. H. 1977. *Climatic requirement.* In G. F. Sprague [editor] *Corn and Corn Improvement.* Agronomy Monograph 18. American Society of Agronomy, Madison, Wisconsin. pp. 591-623.
97. Siddoway, F. H. 1963. *Effects of cropping and tillage methods on dry aggregate soil structure.* Soil Science Society of America Proceedings 27: 452-454.
98. Smika, D. E., and B. W. Greb. 1975. *Nonerodible aggregates and concentration of fats, waxes, and oils in soils as related to wheat straw mulch.* Soil Science Society of America Proceedings 39: 104-107.
99. Soil Conservation Service. 1981. *Land resource regions and major land resource areas of the United States.* Agriculture Handbook No. 296. U.S. Department of Agriculture, Washington, D.C. 156 pp.
100. Soil Conservation Service. 1984. *Conservation technical assistance targeted areas: Cropland/rangeland erosion control.* U.S. Department of Agriculture, Washington, D.C. 1 p.
101. Soil Conservation Society of America. 1982. *Resource conservation glossary* Ankeny, Iowa. 193 pp.
102. Soil Science Society of America. 1987. *Glossary of soil science terms.* Madison, Wisconsin. 44 pp.
103. Soil Survey Staff. 1975. *Soil taxonomy—a basic system*

of classification for making and interpreting soil surveys. Agriculture Handbook No. 436. Soil Conservation Service, U.S. Department of Agriculture, Washington, D.C. 754 pp.

104. Sojka, R. E., G. W. Langdale, and D. L. Karlen. 1984. *Vegetative techniques for reducing water erosion of cropland in the southeastern United States.* Advances in Agronomy 37: 155-181.
105. Swan, J. B., E. C. Schneider, J. F. Moncrief, W. H. Paulson, and A. E. Peterson. 1987. *Estimating corn growth, and grain moisture from air growing degree days and residue cover.* Agronomy Journal 79: 53-60.
106. Trewartha, G. T. 1968. *An introduction to climate.* McGraw Hill Book Co., New York, New York. 408 pp.
107. Unger, P. W. 1984. *Tillage and residue effects on wheat, sorghum, and sunflower grown in rotation.* Soil Science Society of America Journal 48: 885-891.
108. Unger, P. W., and A. F. Wiese. 1979. *Managing irrigated winter wheat residue for water storage and subsequent dryland grain sorghum production.* Soil Science Society of America Journal 43: 582-588.
109. Unger, P. W., G. W. Langdale, and R. I. Papendick. 1988. *Role of crop residues-improving water conservatin and use.* In W. L. Hargrove [editor] *Cropping Strategies for Efficient Use of Water and Nitrogen.* Special Publication 51. American Society of Agronomy, Madison, Wisconsin. pp. 69-100.
110. U.S. Department of Agriculture. 1982. *RCA, a national program for soil and water conservation: 1982.* Final program report and environmental impact statement. Washington, D.C.
111. Van Doren, C. A., and R. S. Stauffer. 1943. *Effect of crop and surface mulches on runoff, soil losses, and soil aggregation.* Soil Science Society of America Proceedings 8: 97-101.
112. Van Doren, D. M., Jr., and D. C. Reicosky. 1987. *Tillage and irrigation.* In J. R. Wilcox [editor] *Soybeans: Improvement, Production, and Uses.* Monograph 16. American Society of Agronomy, Madison, Wisconsin. pp. 391-428.
113. Van Doren, D. M., Jr., and R. R. Allmaras. 1978. *Effect of residue management practices on soil physical environment, microclimate, and plant growth.* In W. R. Oschwald [editor] *Crop Residue Management Systems.* ASA Special Publication 31. American Society of Agronomy, Madison, Wisconsin. pp. 49-83.
114. Voorhees, W. B., R. R. Allmaras, and C. E. Johnson. 1981. *Alleviating temperature stress.* In G. F. Arkin and H. M. Taylor [editors] *Modifying the Root Environment to Reduce Crop Stress.* Monograph 4. American Society of Agricultural Engineers, St. Joseph, Michigan. pp. 217-266.
115. Wagger, M. G., and D. B. Mengel. 1988. *The role of nonleguminous cover crops in the efficient use of water and nitrogen.* In W. L. Hargrove [editor] *Cropping Strategies for Efficient Use of Water and Nitrogen.* ASA Specical Publication No. 51. American Society of Agronomy, Madison, Wisconsin. pp. 115-127.
116. Wells, B. L., T. O. Borich, and J. D. Frus. 1983. *Conservation tillage in an Iowa county.* Journal of Soil and Water Conservation 38: 284-286.
117. Wesley, R. A., and F. T. Cooke. 1988. *Wheat-soybean double-crop systems on clay soil in the Mississippi Valley area.* Journal of Production Agriculture 1: 166-171.
118. Wicks, G. A., and D. E. Smika. 1973. *Chemical fallow in a winter wheat-fallow rotation.* Weed Science 21: 97-102.
119. Wicks, G. A., D. E. Smika, and G. W. Hergert. 1988. *Long-term effects of no-tillage in a winter wheat (*Triticum aestivum*)-sorghum (*Sorghum bicolor*)-fallow rotation.* Weed Science 36: 384-393.
120. Wiese, A. F., editor. 1985. *Weed control in limited-tillage systems.* Monograph 2. Weed Science Society of America, Champaign, Illinois. 297 pp.
121. Williams, D. E., and J. M. Kraft. 1988. *Managing crop residue and tillage pans for pea production.* In 11th International Conference Proceedings, Volume 2. International Soil Tillage Research Organization, Haren, The Netherlands. pp. 927-932.
122. Williams, J. R. 1988. *A stochastic dominance analysis of tillage and crop insurance practices in a semiarid region.* American Journal of Agricultural Economics 70: 112-120.
123. Wilkins, J. R., R. V. Llewelyn, and C. L. Mikesell. 1989. *An economic risk analysis of conservation tillage systems for wheat, grain sorghum, and soybeans in the Great Plains.* Journal of Soil and Water Conservation 44: 234-239.
124. Wischmeier, W. H., and D. D. Smith. 1978. *Predicting rainfall erosion losses: A guide to conservation planning.* Agriculture Handbook No. 537. U.S. Department of Agriculture, Washington, D.C. 58 pp.
125. Wittmus, H. D., D. E. Lane, and B. R. Somerhalder. 1971. *Strip till planting of row crops through surface residue.* Transactions, American Society of Agricutural Engineers 14: 60-63, 68.
126. Woodruff, N. P., and F. H. Siddoway. 1965. *A wind erosion equation.* Soil Science Society of America Proceedings 29: 602-608.
127. Zobeck, T. M., and C. A. Onstad. 1987. *Tillage and rainfall effects on random roughness: A review.* Soil and Tillage Research 9: 1-20.

8

Management of Agricultural Land Receiving Wastewater Sludges

R. H. Dowdy, A. L. Page, and A. C. Chang

Although collection and disposal of water-carried domestic wastes is evidenced in many archeological ruins dating back to antiquity, the systematic treatment of municipal wastewater has been practiced in the western hemisphere for no more than 120 years (*10*). Throughout this period, the sewage collection and treatment system, initially established to combat water-borne epidemic diseases, gradually evolved and today constitutes the backbone of our society's effort to control water pollution.

While the composition of municipal wastewater and the technology used to treat it have become increasingly complicated, the fundamental concept of removing impurities from the spent water prior to its discharge remain unchanged (*11*). Sewage treatment has evolved stepwise from simply removing some suspended solids and pathogens in wastewater (primary treatment); to further pathogen and suspended solids reduction, accompanied by degradation of dissolved oxygen-consuming organic matter (secondary treatment); and, finally, to a system that removes all the above-mentioned pollutants as well as nutrients, such as nitrogen and phosphorus, and a spectrum of potentially toxic chemical constituents resulting from an industrialized society (tertiary treatment).

A modern day "publicly owned treatment work" and its wastewater collection system has numerous service connections and receives discharges from all segments of the service district. All municipal wastewaters contain (a) dissolved-oxygen-demanding organic matter (BOD), (b) disease-causing microorganisms, (c) essential plant nutrients, (d) toxic chemical constituents, and (e) dissolved minerals, which are impurities added during the course of water use. The exact nature of chemical constituents present in the wastewater and their concentrations, to a large extent, reflect the nature of a community (types of industries and volume of industrial discharge, lifestyle and living standard, etc.) (*42, 47*). For a majority of the impurities present in wastewater, the law of conservation of mass dictates that they be retained with either the purified water effluent or the treatment residue (sludge). Some decomposition products, however, do escape the treatment process as gaseous by-products. As the treated effluent becomes progressively "pure" with treatment, the treatment residues (sludges) become more and more concentrated with pollutants and increase in volume. Processing (thickening, digesting, dewatering, and drying) and disposal of the sludges are by far the most difficult and costly operation of a publicly owned treatment work.

Both treated wastewater effluents and sludges have been applied on land since the advent of modern wastewater treatment; today, land application remains a technically viable and cost-effective alternative for their ultimate disposal (*12, 33*). Treated effluents and sludge can be valuable sources of irrigation water as well as nitrogen, phosphorus, and other essential plant nutrients (*18*). Tables 1 and 2 present the generalized concentration range, median, and mean of selected chemical constituents of treated wastewater effluents and municipal sludges in the United States. When treated effluents are applied as irrigation water for crop production at the rate of 1.2 meters per hectare annually, inputs of nitrogen, phosphorus, and trace elements could be substantial (*8*). Applying sludges to satisfy the nitrogen requirements for normal crop growth will add more than adequate amounts of phosphorus for

Table 1. Concentrations of macronutrients and trace elements in wastewater from municipal treatment plants.*

Element	Primary Treatment		Secondary Treatment		Typical
	Range	Median	Range	Median	Concentration†
	mg l^{-1}				
N	20 - 60	35	20 - 60	35	20
P	10 - 145	10	6 - 15	10	10
K	10 - 50	20	10 - 30	20	15
As	<0.005 - 0.03	<0.005	<0.005 - 0.023	<0.005	<0.005
B	<0.1 - 2.5	1.0	<0.1 - 2.5	0.7	1.0
Cd	<0.02 - 6.4	<0.02	<0.005 - 0.15	<0.005	0.005
Cr	<0.05 - 6.8	<0.05	<0.005 - 1.2	0.02	0.025
Cu	<0.02 - 5.9	0.10	<0.006 - 1.3	0.04	0.10
Hg	<0.0001 - 0.125	0.0009	<0.0002 - 0.001	0.0005	0.0009
Mo	<0.001 - 0.02	0.007	0.001 - 0.018	0.007	0.005
Ni	<0.1 - 1.5	<0.1	0.003 - 0.6	0.004	0.02
Pb	<0.02 - 6.0	<0.02	0.003 - 0.35	0.008	0.05
Se	<0.005 - 0.02	<0.005	<0.005 - 0.02	<0.005	<0.005
Zn	<0.02 - 20	0.12	0.004 - 1.2	0.04	0.5

*Based on information presented by Bouwer and Chaney (*6*) and accumulated by the authors from a number of sources.
†From Dowdy and associates (*18*).

Table 2. Concentrations of elements in sewage sludges from treatment plants in the United States.*

Element	Number of POTW†	Range	Median	Mean
		%		
N	191	<0.1 - 17.6	3.3	3.6
P	189	<0.1 - 14.3	2.3	2.5
K	192	0.01 - 2.1	0.3	—
S	28	0.02 - 2.64	1.1	1.1
		mg kg^{-1}		
As	45	0.6 - 26.9	4.79	6.71
Be	45	0.16 - 10.3	0.70	1.62
Cd	45	0.74 - 1622	9.55	68.95
Cr	45	18 - 1676	283.0	429.0
Cu	45	23 - 2764	472.0	592.0
Pb	45	42 - 1693	269.0	368.0
Hg	45	0.23 - 10.2	2.38	2.75
Mo	4	0 - 35.6	20.3	17.7
Ni	45	5.8 - 804	69.8	135.0
Se	45	0.16 - 193	1.3	7.3
Zn	45	101 - 7578	792.0	1594.0

*Nitrogen, P, K and S contents in sludges, expressed as a percentage of dry weight, are from Sommers (*49*); concentrations of other elements, expressed in milligrams per kilogram dry weight, were derived from U.S. EPA (*53*).
†POTW represents publicly owned treatment works.

most plants and introduce significant quantities of trace elements into the soils. Some of the chemical constituents added through land application may be harmful to plants, undesirable to consumers, or act as pollutants of the underlying aquifer. Therefore, caution must be exercised to make certain that the soil's ability to assimilate and attenuate land-applied, potentially harmful chemical constituents will safeguard soil productivity and protect the environment from pollutants (*44*).

Processes by Which Soils Assimilate Chemical Constituents

The primary physical, chemical, and biological processes affecting the fate of chemicals in soil, introduced by land application of wastes, are phase separation, degradation, adsorption, precipitation, crop removal, surface runoff, and leaching. Because soil is a complicated and heterogeneous medium and the above-cited processes take place in a soil simultaneously, the partition of a chemical in various compartments of the soil is most likely determined by the result of all these reactions (*41*). The rates of these reactions are influenced by the landscape position, associated climate, and history of waste application. Soil properties and, therefore, soil's ability to assimilate pollutants are functions of the soil's mineralogy, texture and structure, hydrous oxides of iron and aluminum content, pH and buffering capacity, organic matter content, and oxidation-reduction potential (*43*). Waste applications in the form of treated effluents and sludges introduce sufficient amounts of chemicals that the properties of the receiving soils may undergo both immediate and long-term changes. Therefore, waste properties and amounts applied will greatly affect the outcome. Understanding the far-reaching impacts of applying wastes on land, therefore, requires a thorough knowledge of the soil ecosystem. For filtration, surface runoff, and leaching, hydrodynamics of flow through the porous media must be considered.

Cation Exchange, Adsorption, and Precipitation. Cation exchange, adsorption, and precipitation constitute different chemical processes by which a constituent in the soil solution phase may enter into solid phase (*13, 25*). These processes act to restrict the mobility of soluble chemicals introduced into a soil, thus reducing the chemicals' potential to pollute surface and groundwater and reducing their availability to plants. Because current analytical methodology is limited to the assay of constituents in solution, it is not now possible to directly distinguish among the extents of attenuation attributable to cation exchange, adsorption, and precipitation occurring in a soil (*15*).

Precipitation refers to a chemical process whereby two or more dissolved species combine to form a solid. The possibilities of chemical precipitation in soils are almost unlimited, with progressively larger amounts of chemicals removed from solution as the soil desiccates. Only those species that form sparingly soluble compounds and are taken out of the solution more or less permanently are important in land application of wastes. For example, precipitation of phosphates of cadmium, copper, nickel, lead, and zinc can be an important regulator of metal availability to plants because these compounds are quite insoluble and there is an abundance of phosphates in soils and applied wastes (*39*).

Adsorption denotes the chemical and physical adhesion of substances in a thin layer on the surface of a solid medium. The physical adsorption, a result of the van der Waals universal attraction force, is not chemically specific and is effective with small uncharged particles. The chemical adsorption is caused by the electrostatic forces. The reaction is chemical- and site-specific and, once completed, is often not easily reversible. At the low equilibrium concentrations, the adsorption of almost all chemicals follows the linear adsorption isotherm and the rate of chemical adsorption is constant (*26, 28, 32, 35, 46*). At high equilibrium concentrations, the chemical adsorption by a solid phase usually is either a first-order reaction (the Langmuir adsorption isotherm) or an exponential function (the Freundlich adsorption isotherm). Adsorption of trace metal elements, such as cadmium, copper, nickel, zinc, and mercury, and many nonpolar organic chemicals by soils provides an effective means to attenuate their pollution potential in soils.

During weathering, soil particles may acquire both a negative and/or a positive charge through amorphous substitution within the aluminosilicate mineral structure, from association with or dissociation of protons at the surfaces of soil mineral (mainly the edges), hydrous oxides of iron and aluminum, and carboxyl and phenolic groups of soil organic matter. Even though a soil system is electrostatically neutral, the soil particles usually carry a net negative charge, which makes them reactive to positively charged chemical constituents in the soil solution. At these surface sites, cations in solution may exchange stoichiometrically for those that are attached. If chemical inputs to the soil remain constant, a steady-state exchange equilibrium between cations in the solid phase and those in the solution phase will be reached in reasonably short time. The cation exchange is an important soil chemical reaction for the major cations of the soil solution (e.g., Ca^{+2}, Mg^{+2}, Na^{+}, and K^{+}). However, it has not played a significant role in immobilizing trace metals in soils, despite the fact that trace metals are less available in soils of higher cation exchange capacity (*29*).

The capacity of a soil to exchange cations is influenced by properties and the chemical environment of the soil. The soil texture, mineralogy, and organic matter content exhibit considerable influence on the cation exchange capacity. Generally, fine-textured soils have greater surface area and adsorb more cations than the coarse-textured soils at both cation exchange and specific adsorption sites (*5*). Soils with predominantly 2:1 clay minerals usually have greater cation exchange capacities than soils that are dominated by 1:1 clay minerals. Soil adsorption and cation exchange capacity also increase with their organic matter content since soil organic matter provides charged sites and opportunities for cations to form complexes. Negative charges are created when protons dissociate from a molecule. The adsorption and cation exchange capacity of a soil, therefore, will increase as protons are neutralized or as the soil acidity decreases.

Although soil particles normally carry a net negative charge, their surfaces may also be positively charged. The localized positive charge is created by association of protons on exposed mineral and hydrous oxide surfaces and organic matter and thus permit the soils to adsorb anions at these sites. Whereas mineralogy, hydrous oxides of iron and aluminum, and texture are important factors in determining the extent of anion adsorption in a soil, soil acidity exerts the greatest influence. Anion adsorption of a soil increases as the soil acidity increases. Anions are adsorbed either with or without their water of hydration. Because of the water layer, the charged surface and the hydrated anion form a weak outer-sphere complex (nonspecific adsorption). For nonhydrated anions, a strong inner-sphere complex between the charged surface and the anion will form (specific adsorption). Nitrate, sulfate, and chloride in soils are nonspecifically adsorbed and are susceptible to leaching (*9*). Phosphates, arsenites, selenites, fluoride, chromates, molydates, and other oxyanions are specifically adsorbed and are quite immobile in soils (*5, 31*).

Volatilization, Degradation, Leaching, and Crop Removal. Volatilization and degradation are two processes by which chemicals added to soils through waste application can be removed or eliminated. Via the degradation process, pollutants may be chemically and/or biologically converted, and the end products of the reaction usually are no longer pollutants. The volatilization process transfers chemicals in solution and solid phases to the atmospheric phase. In moist soil, volatilization from dilute aqueous solution is determined by the chemical partitioning coefficient between the water and air, K_w, which is defined as the ratio of the equilibrium concentration in water to its corresponding concentration in air at a given temperature (*50*). This parameter provides a relative measure of a chemical's ability to partition between aqueous and gaseous phases in the soil. A low value of K_w indicates low water solubility and a greater tendency for the compound to become volatilized. Because equilibrium is seldom reached in soil, the partition coefficient of a compound is best determined experimentally (*45*). If experimental data are not available, K_W can be estimated. Mathematically, the value of K_W for an organic chemical is the inverse of its Henry's Law constant; it can be approximated by dividing the water solubility of the compound by its saturated vapor density. As a general rule, the higher the vapor pressure of the compound, the higher its potential for volatilization (*51*). Volatilization of a number of industrial organic compounds, such as chloroform, dichlorobromomethane, 1,2-dichloromethane, tetrachloroethane, 1,1,2-trichloroethane, trichloroethane, chlorobenzene, and toluene, is the major way in which they are lost from soil and for foliar uptake of gaseous pollutants by plants.

Degradation refers to the combined biological, chemical, and photochemical transformation of chemicals. Photochemical reactions, however, are important only when the chemical is deposited on the soil surface and plays a minimal role once the chemical has been incorporated into the soil. Chemical reactions important in the degradation of waste chemicals include oxidative and reductive transformation, hydrolytic nucleophilic transformation, or nonhydrolytic nucleophilic displacement reactions. Biodegradations are enzyme-activated biochemical reactions whereby organic substances are decomposed. Bacteria, actinomycetes, and fungi are the most important organisms involved in the breakdown of organic substances in soils. Although the biochemical pathways of the microbial decomposition process are not always completely defined, microbial metabolism is reasonably well understood (*1*). Biochemical reactions, such as β-oxidation, cleavage of ester linkage, ring hydroxylation, ring cleavage, ester hydrolysis, dehalogenation, and dealkylation, are all common to the biodegradation process in soil. Degradation is, no doubt, the most important process involved in the decontamination (detoxification) of organic pollutants added to the soil (*4, 7, 52*). Ideally, degradation of organic compounds should proceed to completion, releasing the innocuous products, carbon dioxide and water.

Except for synthetic polymers manufactured for their stability and highly insoluble, large molecules, few organic constituents can be considered nonbiodegradable. The rates of organic chemical degradation in soil, however, vary substantially; they depend upon the chemical structure of the compound and the physical, chemical, and biological properties of the soil environment into which it is placed. Environmental conditions conducive to the growth of microorganisms will enhance organic chemical degradation. Temperature, aeration, nutrient supply, moisture, organic matter content, and soil pH all influence the nature of microbial populations in the soil. Degradation of organic chemicals follows first-order reaction kinetics, and the biodegradability of a chemical may be described by its half-life. Under ideal conditions, biodegradation half-lives for simple organic compounds (short-chained fatty acids, sugars, amino acids, etc.) range from hours to days; carbamates, phenols, phthalate esters, and nitrogenous organics range from weeks to months; and monocyclic aromatics, polynuclear aromatics (PAHs), and halogenic biphenyls (PCBs) range from months to years. In soils, those organic chemicals that are resistant to degradation will not be rapidly assimilated into the carbon cycle and, if incorporated into biomass, may be transferred ecologically in a manner that affects organisms at higher tropic levels.

The use of modern biotechnology techniques to decontaminate toxic organic waste is still in its infancy. Using genetically engineered microorganisms in soils is still in the conceptual and experimental stages. Manufactured microorganisms may accelerate degradation and/or carry the process to a more complete state (carbon dioxide and water). Limited findings nevertheless show great promise for biotechnology to increase the rates by which waste chemicals can be decomposed in soil (*24*).

A number of chemical and biochemical pathways can serve to decontaminate inorganic waste chemicals added to soils. Arsenic, mercury, and selenium, for example, can be transformed to volatile forms and released to the atmosphere, where they are diluted to the extent that they are no longer harmful (*27, 36*). Products of degradation can also become sources of other environmental problems. Nitrogenous organic compounds, upon mineralization, produce ammonia, which may accumulate in soils to levels toxic in plants and microorganisms. The dissolved ammonia (ammonium ions), in turn, can be oxidized to nitrate, which is not only an essential plant nutrient but also a potential groundwater pollutant. The interrelation among various degradation pathways, as well as among other chemical reactions, makes it difficult to operate a land application system for which the attenuation of all pollutants can be maximized (*34*).

Leaching involves transporting chemical constituents by mass flow through downward water movement. It occurs when the water application exceeds evapotranspiration. It is the process by which pollutants can be transported to the aquifer. Boron, for example, is an element that is phytotoxic when it is present in soil solutions above certain critical levels (*22*). It is also an element that occurs in soil solutions as a neutral molecule and, consequently, is only weakly adsorbed. Where sludges containing boron are added to soils, the boron can be removed by leaching (*9*). Sludges could add soluble salts to the soil to the extent that plant growth would suffer. In operations in which leaching is employed to remove soluble constituents from a soil zone, care must be taken to guard against groundwater contamination.

Harvesting crops is not only an effective means to remove nitrogen and phosphorus from soils, but also a process that changes the chemical equilibrium in soil solution. In managing land application of nitrogenous wastes, crops play an important role in minimizing nitrate contamination of groundwater. Amounts of nitrogen removed by high-yielding corn (*Zea mays* L.), barley (*Hordeum vulgare* L.), and potato (*Solanum tuberosum* L.) crops often range from 100 to 150 kilograms per hectare per year. Grasses, if harvested, are also highly effective in removing nitrogen from soil. Reed canarygrass and coastal bermudagrass, for example, are capable of removing from 300 to 450 kilograms of nitrogen per hectare per year. Amounts of phosphorus removed by crops, although significant, are less than those of nitrogen and commonly range from 20 to 40 kilograms of phosphorus per hectare. Removal of trace elements and trace organics by crop harvesting is quite small and, consequently, an ineffective process to remove these chemicals following their application to soil.

A Case Study: Rosemount Watershed

All mechanisms and processes controlling the fate of sludge-borne constitutes in the soil occur concurrently when sludge is applied on the land. Observations and measurements made in the field as a result of land application of sludge no doubt reflect the integration of these mechanisms and processes. In the case study that follows, no attempt was made to identify specific mechanisms and processes that occurred when sludge was used for crop production. Instead, data are presented that document the impact of sludge application on crop yields and the fate of sludge-borne plant nutrients and trace metals.

The objectives of the field study were:

► To demonstrate safe, efficient, and practical methods for the application of sewage sludge to land in harmony with agricultural usage while controlling surface and groundwater pollution.

► To collect information on specific practices for land application of sewage sludge so that safe management guidelines can be developed for various soil, crop, and climatic situations in northern climates.

The project was initiated in 1973 and continues to date, with minor treatment changes when suggested by previous results.

Experimental Site and Treatment. Soils on the site formed in the silty surface mantle of wind-deposited silt (loess), 1.5 to 6.0 meters thick, overlaying compacted glacial till (*38*). Major soils on convex hilltops and upper side slopes include the Tallula series (coarse-silty, mixed, mesic Typic Hapludolls). These soils have a naturally dark surface horizon. Subsurface horizons are more red and structural aggregates are better expressed. On side slopes, the dominant soils include the Port Bryon

Table 3. Composition of sewage sludge applied to the terraced areas.

Component	Area: Grass	Corn
Total solids (g l^{-1})	20	29
	($g\ kg^{-1}$ dry sludge)	
C	250	260
Total N	58	53
NH_4 - N	24	22
P	25	26
K	5	4
	($mg\ kg^{-1}$ dry sludge)	
B	33	34
Cd	6	7
Cr	7,640	5,650
Cu	1,030	750
Ni	21	21
Pb	280	350
Zn	805	915

series (fine-silty, mixed, mesic Typic Hapludolls) that have received enough run-on water to form a thick soil with a subsurface horizon that has greater clay content than the parent material. The major soils in the lower drainageways include the Colo series (fine-silty, mixed, mesic Cumulic Haplaquolls) with a surface horizon up to 1.5 meters thick. They are poorly drained soils with a perched water table at a depth of 1.0 meter in most years. Surface-soil pH ranged from 6.5 to 7.3, initially and has remained unchanged throughout the study.

Water entering the soil percolates to silt-till contacts and remains perched on the dense till. This water may move laterally downslope, but is unable to move directly into the deep (35 to 45 meters) groundwater aquifer (St. Peter sandstone).

The experimental site consists of parallel terraces constructed on a 16-hectare watershed; slopes range from 2 to 10 percent (Figure 1). Larson and associates (*40*) have detailed the engineering design of the watershed as prepared by the U.S. Soil Conservation Service. Terrace ridges were spaced 40 meters apart, and terrace channels were graded to perforated PVC inlet pipes placed in natural depressions. Solid PVC pipe connect each inlet pipe with individual sampling stations where runoff flow rate was measured and runoff was sampled automatically.

The watershed was divided into 10 separate areas, five cropped to continuous corn and five to reed canarygrass sod (*Phaloris arendenacea* L.). In 1984, the grass crops were terminated because of weed competition. The entire watershed is now cropped to continuous corn. Prior to 1984, four areas (terrace intervals) of each crop were treated as replicated blocks and received mean annual sludge applications of 12 and 15 megagrams per hectare, respectively. Another area for each crop received no sludge, but received adequate mineral fertilizer for optimal crop growth. Details of crop management were presented by Clapp and associates (*14*) and Duncomb and associates (*21*).

Metal concentrations in the sewage sludge were relatively low (Table 3) and, generally, below median concentrations reported for sludges in the North Central Region of the United States (*18*). Chromium concentrations in the sludge were high and variable. Because chromium is insoluble in soil systems, it served as an excellent tracer for following particulate erosion from the watershed.

For corn, the liquid sludge was injected into the soil each fall after harvesting the crop. The grass areas were irrigated with liquid sludge each spring and after each of three cuttings (before significant regrowth) during the summer and early fall. In addition, one grass area received additional sludge applications during the winter. Liquid sludge was spread by tanker wagon onto frozen, snow-covered ground.

The 2-molar hydrochloric acid sewage-sludge ash (450 ° C) extracts were analyzed for cadmium, chromium, copper, nickel, lead, and zinc by atomic adsorption spectroscopy using continuous-source deuterium-lamp background correction (*20*). Total nitrogen and ammonium nitrate were determined in liquid samples by micro-Kjeldahl procedures involving digestion, distillation, and titration (*14*). Total carbon was measured by dry combustion. Samples for total phosphorus and potassium were dry-ashed at 450 ° C for 24 hours, equilibrated with 2-molar hydrochloric acid and the supernant analyzed by inductively coupled plasma-emission spectroscopy (ICP).

Elemental composition of plant tissue was

determined in a manner similar to that for sewage sludge. Soil and sediment samples were equilibrated overnight with 1-molar nitric acid, and the supernant was analyzed spectroscopically. Water samples were acidified (pH < 1), concentrated, dried, ashed at 450 °C, extracted with 2-molar hydrochloric acid, and analyzed for phosphorus, calcium, sodium, and trace metals spectroscopically. For water samples, nitrate nitrogen and phosphate phosphorus were determined colorimetrically by an autoanalyzer (*14*).

It was not possible to analyze the data statistically because of lack of true replication of the control area. Because trace metal concentrations in runoff waters were low, dif-

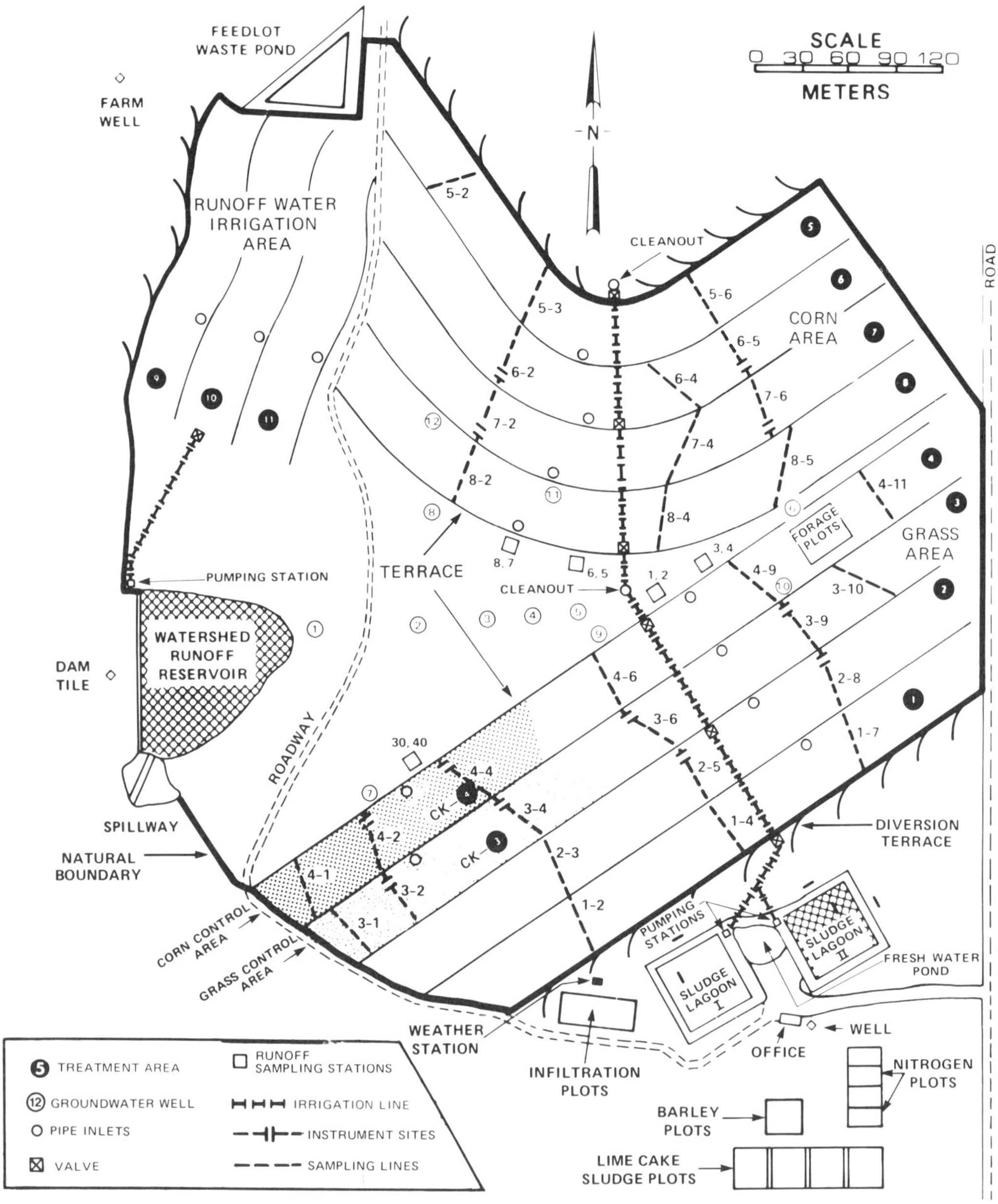

Figure 1. Rosemount sewage sludge watershed (engineering design and sample collection locations).

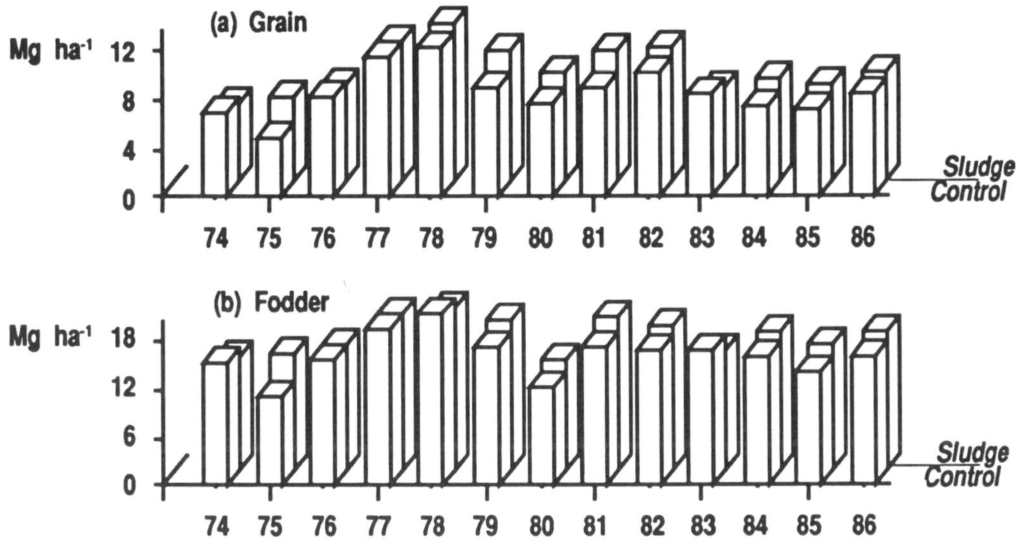

Figure 2. Corn dry matter production (1974-1986) at the Rosemount watershed.

ferences of less than a factor of two are not considered meaningful.

Results and discussion

Plant Growth. Corn and grass yields were excellent for the Minnesota climate, with average dry matter production slightly, but consistently, greater for the sludge treatments than for comparably fertilized control treatments (*38*). Dry matter production for sludge and fertilized control treatments followed the same year to year trends, suggesting that external, rather than treatment, factors contributed greatly to yield variations over years. Corn grain yields (15.5 percent moisture) averaged 8.9 (1.9 standard deviation) and 8.2 (1.9) megagrams per hectare for sludge and control treatments, respectively (Figure 2). Comparable corn dry matter (fodder) yields were 16.5 (2.1) and 15.8 (2.6) megagrams per hectare for sludge and control treatments, respectively. Reed canarygrass yields (Figure 3) followed the same general trends that were observed for corn, with mean dry matter production of 11.0 (1.3) and 9.6 (1.4) megagrams per hectare for sludge and control treatments, respectively. Therefore, land application of sewage sludge is compatible with high levels of crop production.

Nutrients. Macronutrient uptake by corn was essentially the same from the sludge source versus commercial, mineral fertilizer treatments (Figure 4). However, corn removed only an equivalent of 35 percent of the sludge-borne nitrogen. During 1983, sludge-nitrogen application rates were brought into "balance" with corn uptake to correct this disparity between nitrogen application and nitrogen uptake. Although only an equivalent of 13 percent of the sludge-borne phosphorus was taken up by corn, phosphorus does not have the leaching potential associated with nitrogen. Corn mined native soil potassium, but potassium deficiency was never observed. Nutrient uptake patterns for reed canarygrass were very similar to those observed for corn (Figure 5). However, on a mean annual basis, reed canarygrass accumulated 40 percent more sludge-borne nitrogen and 15 percent more sludge-borne phosphorus than corn accumulated. These data clearly demonstrate the importance of balancing sludge applications with the nutrient requirements of the crop.

Water. Sludge-borne nitrogen applications in excess of crop needs (Figure 4) lead to elevated nitrate nitrogen concentrations in soil water during year 5 (Table 4). The very high nitrate nitrogen levels in soil water at 1.5 meters, extracted from both the control and sludge-treated corn areas, probably represents an accumulated pulse of 5 years of overfertilization with nitrogen in both treatments. Some of the sludge-borne nitrate-nitrogen was also apparent in perched water, where concentrations of 8 versus 60 milligrams nitrate nitrogen per liter for the control and sludge treatments (corn), respectively, were observed. Reed canarygrass was much more effective in using applied nitrogen.

By year 15, after nitrogen application had been brought into closer balance with crop use, nitrate nitrogen concentrations in perched water were not enhanced as a result of sludge application (Table 4). Nitrate contamination of groundwater was never observed.

Sludge-borne phosphorus was never detected in soil, perched, or groundwater (Table 4).

Trace Metals. Total sludge and metal loadings through years 5, 10, and 13 are listed in table 5. With the exception of chromium, which was applied at up to 1,180 kilograms per hectare over the 10-year period, sludge-borne metal loadings were relatively low.

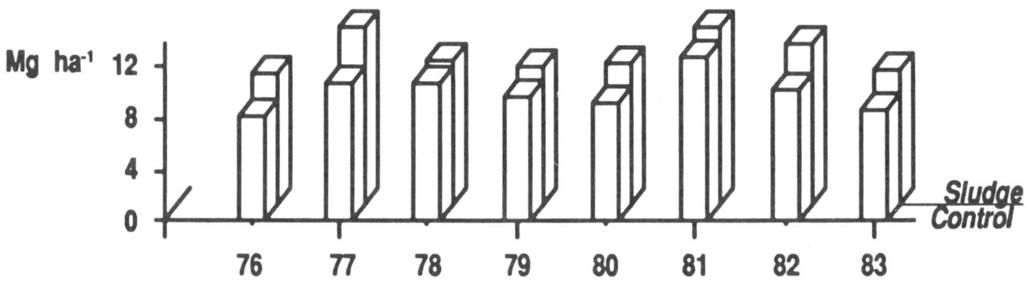

Figure 3. Dry matter production of reed canarygrass (1976-1983) at the Rosemount watershed.

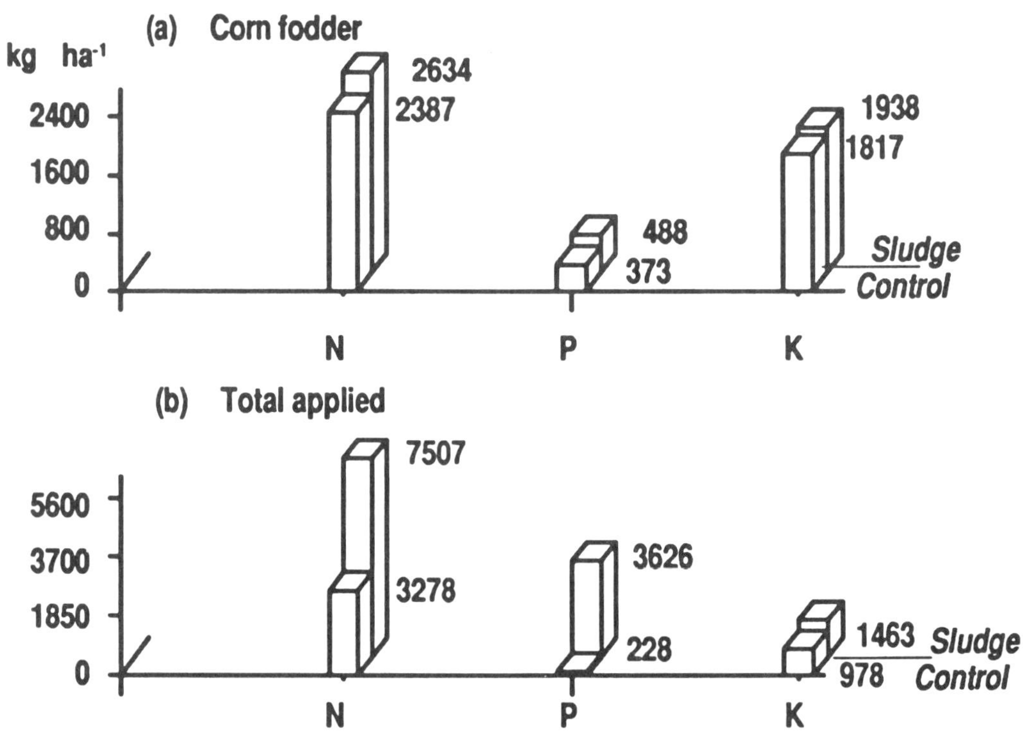

Figure 4. Uptake of nitrogen, phophorus, and potassium by corn (a) versus amount applied (b) at the Rosemount watershed (1974-1986).

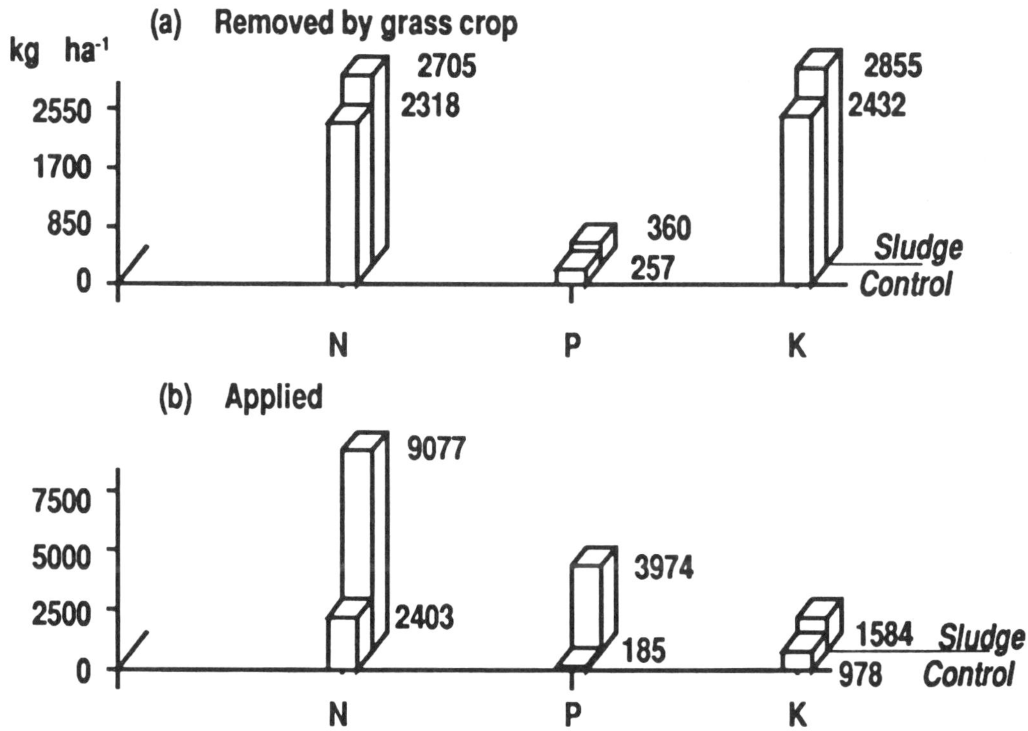

Figure 5. Uptake of nitrogen, phosphorus, and potassium by reed canarygrass (a) versus amount applied (b) at the Rosemount watershed (1976-1983).

Plant Tissues. Zinc was the only sludge-borne trace metal that increased in corn tissues as a result of sludge additions (Table 6). Concentrations of zinc in stover increased from 17 (control) to 51 milligrams per kilogram as a result of 13 years of sludge additions (128 kilograms of zinc per hectare). Although elevated (51 milligrams of zinc per kilogram tissue), this zinc concentration is relatively low when compared with situations in which higher zinc loadings have occurred (*3*). The increased zinc was not translocated into grain.

The apparent increased metal concentrations in reed canarygrass (Table 6) grown on sludge-treated areas were probably an artifact of the sampling scheme. It is suggested that tissues were surface contaminated with sludge by rain splash. Although sludge was not applied to the growing tissue, it was applied on the soil surface and remained in that position as a mat of residue. This suggestion is based, in part, on the high chromium levels observed for the sludge treatment and the fact that plants do not readily take up chromium (*2*). Plant samples were not washed prior to analysis.

Runoff Water. Metal concentrations (Table 7) in runoff water were low (micrograms per liter). Attention is directed to snowmelt and early spring runoff when the soil surface is most vulnerable to erosion and to runoff events greater than 0.01 meter. Collectively, the data focus on a worst possible scenario (*16*). Cadmium, nickel, and lead concentrations in runoff were not affected by sludge applications. Copper levels increased as a result of sludge applications. Maximum copper concentrations (approximately 100 micrograms per liter) were observed in years following additional sludge applications on frozen, snow-covered surfaces. For year 5, total copper runoff from grass areas that received a winter sludge application was 0.17 kilogram per hectare compared to 19.5 kilograms per hectare of copper applied the previous year. To further place these copper levels in perspective, U.S. Public Health Service drinking water standards are 1,000

Table 4. Nutrient (NO_3-N and PO_4-P) concentrations in soil water, perched water, and groundwater after 5, 10, and 15 years of annual sludge applications.

		Year 5*				Year 10†		Year 15†	
		Soil Water Depth (m)							
Nutrient	Treatment	0.6	1.5	Perched Water‡	Groundwater§	Perched Water	Groundwater	Perched Water	Groundwater
		mg l⁻¹							
					Grass				
NO_3-N	Control	20	10	8	0.5	52	1.0	38	0.7
	Sludge	102	70	56	NA#	50	NA	34	NA
PO_4-P	Control	<0.01	<0.01	0.01	0.02	<0.01	<0.01	<0.02	<0.02
	Sludge	0.08	0.01	0.02	NA	<0.01	NA	NA	NA
					Corn				
NO_3-N	Control	90	230	8	0.5	52	1.0	38	0.7
	Sludge	160	350	60	NA	88	NA	46	NA
PO_4-P	Control	0.04	0.03	0.01	0.02	<0.01	<0.01	<0.02	<0.02
	Sludge	0.01	0.02	0.01	NA	<0.01	NA	<0.02	NA

*Samples taken monthly from May through October.
†Samples taken biannually (April and November).
‡Samples from saturated zone, 0.5 to 1.5 m from soil surface above dense glacial till.
§Samples from groundwater aquifer, approximately 40 m under watershed.
#Not applicable.

Table 5. Cumulative sludge and trace metal loadings on terraced areas.

Year	Sludge Applied	Trace Metal Zn	Cu	Cd	Ni	Pb	Cr
	(Mg ha⁻¹)*	kg ha⁻¹					
				Grass			
5	58	40	35	0.3	1.1	15	175
10	160	120	150	0.9	3.4	50	1180
				Corn			
5	52	50	40	0.3	0.9	15	195
10	130	120	100	0.8	2.8	45	750
13	160	130	105	0.9	3.4	50	905

*Expressed on a dry weight basis, 105 degrees Celsius.

Table 6. Mean concentrations of trace metals in corn (13 years) and reed canarygrass (10 years) grown on terraced areas receiving annual sludge applications.

Tissue	Treatment	Trace Metal Zn	Cu	Cd	Ni	Pb	Cr
		mg kg⁻¹					
Corn Stover	Control	17	7.4	0.16	0.6	1.1	1.1
	Sludge	51	7.5	0.27	0.9	1.2	1.9
Corn Grain	Control	23	2.8	0.19	0.3	<0.4	<0.2
	Sludge	31	2.2	0.21	<0.3	<0.4	<0.2
Grass	Control	27	6.0	0.06	2.1	0.8	2.5
	Sludge	63	14	0.11	2.0	1.6	21.5*

*Probably external surface contamination due to rain splash and does not represent biological uptake.

Table 7. Elemental content of runoff water from terraced areas after 5 and 10 years of annual sludge applications.

Year	*Date*	*Treatment*	*Runoff (m)*	*Zn*	*Cu*	*Cd*	*Ni*	*Pb*	*Cr*
				$\mu g\ l^{-1}$					
				Grass					
5	Snowmelt	Control	0.022	31	7	<2	<10	<15	6
		Sludge	0.029	25	32	<2	<10	<15	12
		Sludge*	0.030	30	40	<2	<10	<15	23
	4/07	Control	0.022	46	20	<2	12	<12	3
		Sludge	0.028	57	74	<2	9	<12	25
		Sludge*	0.031	52	92	<2	14	<12	26
10	Snowmelt	Control	0.006	7	1	<1	<2	<15	4
		Sludge	0.005	13	29	<1	6	<15	4
		Sludge*	0.006	15	37	<1	7	<15	6
	5/09	Control	0.011	10	8	<1	<3	<15	10
		Sludge	0.014	20	48	<1	9	<15	18
		Sludge*	0.031	27	44	<1	10	<15	26
				Corn					
5	Snowmelt	Control	0.021	22	10	<1	<10	<15	11
		Sludge	0.036	46	72	<1	<10	<15	33
	4/07	Control	0.021	19	10	<2	4	<12	3
		Sludge	0.039	27	18	<2	9	<12	5
10	Snowmelt	Control	—	16	4	<1	4	<15	26
		Sludge	—	38	10	<1	<3	<15	21
	5/09	Control	0.028	10	9	<1	5	<15	15
		Sludge	0.031	20	11	<1	5	<15	16

*Received winter applications of 4.0 and 5.3 megagrams per hectare sludge in January of years 5 and 6, but no further winter applications.

micrograms per liter. The enhanced copper levels in runoff that resulted from sludge applications are similar to those noted by Hinesly and Jones (*30*). The prominence of copper in runoff probably resulted from the stable complexes it forms with soluble organic materials of sludge and/or grass residue origin. Copper enrichment of runoff from sludge-treated corn areas was less pronounced.

Elevated zinc concentrations were noted in runoff from sludge-treated areas, although the evidence is far less convincing than that observed for copper. Runoff chromium concentrations tended to be slightly higher in stormwater from grass areas. This may result in an association between chromium and eroded colloidal particulates. Greatest chromium losses were associated with eroded sediments (data not shown) from corn areas, where sediments from early spring storms contained 10 times as much chromium as sediments from control areas. Sediments from these areas often contained 400 to 500 milligrams of chromium per kilogram following 10 years of sludge applications (750 kilograms of chromium per hectare). However, sediment runoff was minimal due to the terrace system; hence, only a very small fraction of applied chromium left the landscape.

Soil Water. The evidence (Table 8) for trace metal migration, even to a 0.6-meter depth, was very weak (*16*). A hint of chromium and nickel movement was suggested, but no evidence of movement was detected in year 6 (data not shown). Consistently higher concentrations of sodium and calcium were observed in soil water extracted from sludge treated areas. Sludge-borne trace metal enrichment of perched or groundwater was never observed.

Soil. Elevated concentrations of chromium, copper, lead, and zinc (Table 9) were observed in the 0.08- to 0.15-meter soil layer beneath grass areas following 10 years of surface sludge applications (*17*). Because of massive loadings (1,180 kilograms of chromium per hectare), a 30-fold increase in chromium concentration was noted: 155 and 5 milligrams

Table 8. Elemental concentration of soil water from terraced areas after 5 years of annual sludge applications.

Treatment	*Depth (m)*	*Zn*	*Cu*	*Cd*	*Ni*	*Pb*	*Cr*	*Na*	*Ca*
	(m)	*μg l⁻¹*						*mg l⁻¹*	
Grass Control	0.6	95	18	<1	6	<12	<7	5	75
Grass Sludge	0.6	110	39	<1	13	<12	<7	70	225
Corn Control	0.6	54	4	<1	6	<12	<7	7	65
Corn Sludge	0.6	61	9	<1	12	<12	<7	79	206
Grass Control	1.5	260	31	<1	5	<12	<7	8	74
Grass Sludge	1.5	80	23	<1	8	<12	<7	27	207
Corn Control	1.5	68	4	<1	9	<12	<7	17	164
Corn Sludge	1.5	74	11	<1	31	<12	<7	40	309

Table 9. Extractable (1.0-molar HNO_3) metals as a function of soil depth from grass terraces after 10 years of annual sludge applications.

Depth (m)	*Zn*	*Cu*	*Cd*	*Ni*	*Pb*	*Cr*	*P*	*Na*	*Ca*
	mg kg⁻¹						*g kg⁻¹*		
				Control					
0-0.07	12	6	0.18	4	10	20	0.2	0.0	2.1
	(3)*	(2)	(0.02)	(1)	(1)	(20)	(0.1)	(0.0)	(0.5)
0.08-0.15	7	4	0.12	4	7	5	0.1	0.0	3.2
	(1)	(0)	(0.02)	(1)	(0)	(0)	(0.0)	(0.0)	(0.6)
0.16-0.30	7	4	0.05	4	7	5	0.1	0.0	3.2
	(1)	(0)	(0.00)	(1)	(1)	(0)	(0.0)	(0.0)	(0.5)
0.31-0.60	9	5	0.05	6	8	5	0.1	0.0	3.4
	(2)	(0)	(0.00)	(0)	(0)	(0)	(0.0)	(0.0)	(0.3)
				Sludge					
0-0.07	58	68	0.71	6	28	345	1.1	0.1	5.0
	(17)	(23)	(0.11)	(1)	(6)	(125)	(0.3)	(0.0)	(0.6)
0.08-0.15	34	37	0.32	5	18	155	0.6	0.1	4.8
	(7)	(9)	(0.02)	(0)	(3)	(40)	(0.1)	(0.0)	(0.4)
0.16-0.30	23	25	0.18	5	13	110	0.4	0.1	4.8
	(6)	(8)	(0.03)	(1)	(2)	(40)	(0.1)	(0.0)	(0.5)
0.31-0.60	11	7	0.09	7	8	15	0.1	0.1	4.6
	(1)	(1)	(0.01)	(1)	(0)	(5)	(0.0)	(0.0)	(0.6)

*Standard error means are listed in parentheses.

Table 10. Extractable (1.0-molar HNO_3) metals as a function of soil depth from corn terraces after 10 years of annual sludge applications.

Depth (m)	*Zn*	*Cu*	*Cd*	*Ni*	*Pb*	*Cr*	*P*	*Na*	*Ca*
	mg kg⁻¹						*g kg⁻¹*		
				Control					
0-0.15	8	3	0.10	4	8	5	0.1	0.0	3.4
	(1)*	(0)	(0.00)	(1)	(0)	(0)	(0.1)	(0.0)	(0.5)
0.16-0.30	8	5	0.10	5	12	5	0.1	0.0	4.1
	(1)	(2)	(0.03)	(1)	(1)	(0)	(0.0)	(0.0)	(0.4)
0.31-0.60	7	4	0.08	5	7	5	0.0	0.0	3.1
	(1)	(1)	(0.03)	(1)	(1)	(0)	(0.0)	(0.0)	(0.5)
				Sludge					
0-0.15	46	52	0.33	6	24	220	1.0	0.1	8.9
	(2)	(2)	(0.02)	(0)	(1)	(10)	(0.0)	(0.0)	(1.6)
0.16-0.30	27	26	0.26	6	15	100	0.6	0.1	12.7
	(3)	(3)	(0.03)	(0)	(1)	(10)	(0.0)	(0.0)	(3.1)
0.31-0.60	10	5	0.07	6	6	5	0.2	0.1	18.9
	(1)	(0)	(0.01)	(0)	(0)	(0)	(0.0)	(0.0)	(4.9)

*Standard error means are listed in parentheses.

of chromium per kilogram soil for sludge treated versus control areas, respectively. This trend continued with profile depth. It is possible that sludge-borne chromium remained in a relatively more soluble organic complex once applied and, hence, was subject to movement through macropores with percolating water into the soil. However, no evidence for chromium movement was detected in soil water samples taken at 0.6 meter (*16*).

Enhanced levels of zinc and copper were not observed below 0.3 meter. Both zinc and copper concentrations increased from less than 10 milligrams per kilogram to 24 milligrams per kilogram in the 0.16- to 0.3-meter layer as a result of 10 years of sludge application (120 and 150 kilograms per hectare of zinc and copper, respectively). A hint of elevated copper concentrations in soil water at 0.6 meter from this sludge-treated area has been reported (*16*). Williams and associates (*54*) noted movement of zinc and copper into the 0.25- to 0.3-meter soil layer following 8 years of sludge incorporation (5,257 and 880 kilograms per hectare of zinc and copper, respectively) into the top 0.2 meter of soil. But Emerich and associates (*23*) did not report copper movement below the sludge-soil interface. Kelling and associates (*37*) noted that copper did not move in sludge-amended soil, but zinc concentrations did increase at depths less than 0.3 meter.

Cadmium concentrations (Table 9) in soil down to the 0.16- to 0.3 meter layer increased as a result of sludge applications. This increase was only 0.1 milligram per hectare, as expected by the small cadmium loading (0.9 kilogram per hectare). Some cadmium movement into soil below the incorporation zone has been reported (*54*), but is in contrast to much of the published literature (*19*). A hint of lead and calcium movement into the 0.08- to 0.15-meter zone was suggested; however, nickel migration into soil did not occur, although only 34 kilograms per hectare were applied.

Elevated concentrations of trace metals (Table 10) in the 0- to 0.15-meter soil layer for terraced areas cropped to continuous corn resulted from sludge incorporation into this zone by tillage (*17*). In contrast to sludge-treated grass areas, it appears that chromium did not move as readily in the area cropped to corn. This lesser movement may be related to a lower total chromium loading, but may also be the result of faster mineralization of sludge-borne organo-chromium complexes that, when incorporated into the soil, leave chromium in a less mobile state. Zinc, copper, and cadmium did not move below 0.3 meter, but some movement of these metals into the 0.16- to 0.3-meter zone did occur, where concentrations of zinc, copper, and cadmium of 27, 26, and 0.26 milligrams per hectare soil, respectively, were observed. Because these increases were at least three times greater than values for comparable control areas, the higher values cannot be explained entirely as an artifact of inaccurate measurement of tillage depth.

Sludge-borne calcium and sodium were observed to move into the 0.31- to 0.6-meter soil layer in areas cropped to corn. Although the processes by which calcium migrated are not understood, the potential for using sludge to ameliorate subsoil acidity is suggested.

In Summary

We have discussed many of the mechanisms and processes that occur concurrently in soil receiving applications of wastewater sludge. While unable to document the relative importance of specific mechanisms and processes that are operative in a field situation, the Rosemount watershed case study demonstrates that sewage sludge can be used for crop production with little impact on environmental quality. Over the 10-year period, corn and reed canarygrass production was excellent, with no significant differences in yields between fertilized control areas and sludge-treated areas. While sludge-borne cadmium, nickel, and lead were not detected in runoff waters, enhanced copper concentrations in runoff were noted, but they remained 10 times below the U.S. Public Health Service drinking water standards. Slightly elevated zinc concentrations in runoff from sludge-treated areas were noted occasionally, whereas chromium losses were associated with eroded sediment. Evidence of sludge-borne trace metals in soil water was very weak, whereas sodium and calcium enrichment of soil water at 1.5 meters was consistent. Nitric acid extracts of soil samples suggested that sludge-borne chromium, copper, and zinc had migrated to a depth of 0.3 meter following 10 years of sludge application. Movement of trace metals below 0.3 meter was not evident. Nitrate content of soil water at 1.5 meters was elevated as a result of excessive sludge applications during the early years of the study.

REFERENCES CITED

1. Alexander, M. 1981. *Biodegradation of chemicals of environmental concern.* Science 211: 132-138.
2. Allaway, W. H. 1977. *Food chain aspects of the use of organic residues.* In L. F. Elliott and F. J. Stevenson [editors] *Soils for Management of Organic Wastes and Wastewaters.* Soil Science Society of America, Madison, Wisconsin.
3. Bidwell, A. M., and R. H. Dowdy. 1987. *Cadmium and zinc availability to corn following termination of sewage sludge applications.* Journal of Environmental Quality 16: 438-442.
4. Boethling, R. S., and M. Alexander. 1979. *Microbial degradation of organic compounds at trace levels.* Environonmental Science and Technology 13(8): 989-991.
5. Bohn, H., B. McNeal, and G. A. O'Connor. 1985. *Soil chemistry.* John Wiley and Sons, New York, New York.
6. Bouwer, H., and R. L. Chaney. 1974. *Land treatment of wastewater.* Advances in Agronomy 26: 133-176.
7. Bunch, R. L., and C. W. Chambers 1967. *A biodegradability test for organic compounds.* Journal of Water Pollution Control Federation 39(2): 181-187.
8. Chang, A. C., and A. L. Page. 1977. *Trace elements in wastewater.* California Agriculture 31(5): 31-32.
9. Chang, A. C., A. L. Page, P. F. Pratt, and J. E. Warneke. 1988. *Leaching of nitrate from freely-drained irrigated fields treated with municipal sludges.* In D. R. Hay [editor] *Proceedings, 1988 National Conference on Irrigation and Drainage Engineering.* American Society of Agricultural Engineers, St. Joseph, Michigan. pp. 455-467.
10. Chae, E. S. 1964. *Nine decades of sanitary engineering, Part II.* The awakening. Water Work Waste Engineering 34(b): 58-59, 86.
11. Chase, E. S. 1964. *Nine decades of sanitary engineering, Part III—Back to the land.* Water Work Waste Engineering 34(7): 48-49, 79.
12. Chase, E. S. 1964. *Nine decades of sanitary engineering, Part IV—Science enters.* Water Work Waste Engineering 34(8): 49-50, 78.
13. Chiou, C. T., L. J. Peters, and V. H. Freed. 1979. *A physical concept of soil-water equilibria for nonionic compounds.* Science 206: 831-832.
14. Clapp, C. E., D. R. Duncomb, W. E. Larson, D. R. Linden, R. H. Dowdy, and R. E. Larson. 1977. *Crop yields and water quality after application of sewage sludge to an agricultural watershed.* In R. C. Loehr [editor] *Proceedings, 1977 Cornell Agricultural Waste Management Conference.* Ann Arbor Science Publishers, Inc., Ann Arbor, Michigan. pp. 185-198.
15. Corey, R. B. 1981. *Adsorption vs. precipitation.* In M. A. Anderson and A. J. Rubin [editors] *Adsorption of Inorganics at Solid-Liquid Interfaces.* Ann Arbor Science Publishers, Inc., Ann Arbor, Michigan. pp. 161-182.
16. Dowdy, R. H., A. M. Bidwell, and C. E. Clapp. 1987. *Trace metal movement in an agricultural watershed.* In *Proceedings, 6th Internationl Conference, Heavy Metals in Environment.* CEP Consultants, Edinburgh, Scotland. pp. 384-386.
17. Dowdy, R. H., M. S. Dolan, C. E. Clapp, and W. E. Larson. 1989. *Trace metal distributions in a loess soil following 10 years of sludge applications.* In *Proceedings, Seventh International Conference, Heavy Metals in the Environment.* CEP Consultants, Edinburgh, Scotland.
18. Dowdy, R. H., R. E. Larson, and E. Epstein. 1976. *Sewage sludge and effluent use in agriculture.* In *Land Application of Wastewater Materials.* Soil Conservation Society of America, Ankeny, Iowa. pp. 138-153.
19. Dowdy, R. H., and V. V. Volk. 1983. *Movement of heavy metals in soils.* In D. W. Nelson, K. K. Tanji, and D. E. Elrich [editors] *Chemical Mobility and Reactivity in Soil Systems.* Special Publication No. 11. Soil Scciety of America, Madison, Wisconsin. pp. 229-240.
20. Dowdy, R. H., W. E. Larson, J. M. Titrud, and J. J. Latterell. 1978. *Growth and metal uptake of snap beans grown on sewage sludge amended soil: A four-year study.* Journal of Environmental Quality 7: 252-257.
21. Duncomb, D. R., W. E. Larson, C. E. Clapp, R. H. Dowdy, D. R. Linden, and W. K. Johnson. 1982. *The effect of liquid sewage sludge application on crop yield and water quality in a closed watershed.* Journal of Water Pollution Control Federation 54: 1,185-1,193.
22. Eaton, F. M. 1944. *Deficiency, toxicity and accumulation of boron in plants.* Journal of Agriculture Research 69: 237-277.
23. Emmerich, W. E., L. J. Lund, A. L. Page, and A. C. Chang. 1982. *Solid phase forms of heavy metals in sewage sludge-treated soils.* Journal of Environmental Quality 11: 178-181.
24. Focht, D. D. 1988. *Performance of biodegradative microorganisms in soil: Xenobiotic chemicals as unexploited metabolic niches.* In G. Omenn [editor] *Environmental Biotechnology.* Plenum Press, New York, New York pp. 15-30.
25. Fuller, W. H. 1977. *Movement of selected metals, asbestos, and cyanide in soil: Applications to waste disposal problems.* EPA-600/2-77-020. U.S. Environmental Protection Agency, Cincinnati, Ohio.
26. Garcia-Miragaya, J., and A. L. Page. 1978. *Sorption of trace quantities of cadmium by soils with different chemical and mineralogical composition.* Water, Air, and Soil Pollution 9: 289-299.
27. Gavis, J., and J. F. Ferguson. 1972. *The cycling of mercury through the environment.* Water Research 6: 989-1,008.
28. Griffin, R. A., and E.S.K. Chien. 1980. *Attentuation of water soluble polychlorinated biphenyls by earth material.* EPA-600/2-80-027. U.S. Environmental Protection Agency, Cincinnati, Ohio.
29. Haghiri, F. 1974. *Plant uptake of cadmium as influenced by cation exchange capacity, organic matter, zinc, and soil temperature.* Journal of Environmental Quality 3: 180-183.
30. Hinesley, T. D., and R. L. Jones. 1976. *Heavy metal content in runoff and drainage waters from sludge-treated field lysimeter plots.* In *Proceedings, National Conference on Disposal Residues on Land.* Information Transfer

Inc., Rockville, Maryland. pp. 27-44.

31. Jarrell, W., A. L. Page, and A. A. Elseewi. 1980. *Molybdenum residues in the environment.* Residue Review 74: 1-43.
32. Jenne, E. A. 1968. *Controls on Mn, Fe, Co, Ni, Cu, and Zn concentrations in soils and waters: The significant role of hydrous Mn and Fe oxides.* In R.J.F. Gould [editor] *Advances in Chemistry Series.* Publication No. 73. American Chemical Society, Washington, D.C. pp. 337-387.
33. Jewell, W. J., and B. L. Seabrook. 1979. *A history of land application as a treatment alternative.* Technical Report EPA 430-9-79-012. U.S. Environmental Protection Agency, Washington, D.C.
34. Jury, W. A., W. F. Spencer, and W. J. Farmer. 1983. *Use of models for assessing relative volatility, mobility and persistence of pesticides and other trace organics in soil systems.* In *Hazard Assessment of Chemicals: Current Development.* Academic Press, New York, New York. pp. 11-43.
35. Karickhoff, S. W., D. S. Brown, and T. A. Scott. 1979. *Sorption of hydrophobic pollutants on natural sediments.* Water Research 13: 241-248.
36. Karlson, U., and W. T. Frankenberger, Jr. 1989. *Accelerated rates of selenium volatilization from California soils.* Soil Science Society of America Journal 53: 749-753.
37. Kelling, K. A., D. R. Keeney, L. M. Walsh, and J. A. Ryan. 1977. *A field study of the agricultural use of sewage sludge: III.* Effects on uptake and extractability of sludge-borne metals. Journal of Environmental Quality 6: 352-358.
38. Knuteson, James, C. E. Clapp, R. H. Dowdy, and W. E. Larson. 1988. *Sewage sludge management: Land application of municipal sewage sludge to a terraced watershed in Minnesota.* Miscellaneous Publication 56-1988. Minnesota Agricultural Experiment Station, University of Minnesota, St. Paul.
39. Korte, N. E., J. Skopp, W. H. Fuller, E. E. Niebla, and B. A. Alesii. 1976. *Trace element movement in soils: Influence of soil physical and chemical properties.* Soil Science 122: 350-359.
40. Larson, R. E., J. A. Jeffery, W. E. Larson, and D. R. Duncomb. 1978. *A closed watershed for applying municipal sludge on crops.* Transactions, American Society of Agricultural Engineers 21: 124-128.
41. Laskowski, D. A., C.A.F. Goring, P. J. McCall, and R. L. Swann. 1982. *Terrestrial environment.* In R. A. Conway [editor] *Environmental Risk Analysis for Chemicals.* Van Nostrand Reinhold Co., New York, New York. pp. 198-240.
42. Levins, P., J. Adams, P. Brenner, S. Coons, G. Harris, C. Jones, K. Thrun, and A. Wechsler. 1979. *Sources of toxic pollutants found in influents to sewage treatment plants: VI.* EPA 68-01-3857. U.S. Environmental Protection Agency, Cincinnati, Ohio.
43. Lindsay, W. L. 1973. *Inorganic reactions of sewage wastes with soils.* In *Recycling Municipal Sludges and Effluents on Land.* National Assocication of State Universities and Land Grant Colleges, Washington, D.C. pp. 91-97.
44. Lisk, D. J. 1972. *Trace metals in soils, plants, and animals.* Advances in Agronomy 24: 267-311.
45. Mackey, D., and P. J. Leinonen. 1975. *Rate of evaporation of low solubility contaminants from water bodies to atmosphere.* Environmental Science and Technology 9(10): 1,178-1,180.
46. Means, J. C., S. G. Wood, J. J. Hassett, and W. L. Banwart. 1982. *Sorption of amino- and carboxy-substituted polynuclear aromatic hydrocarbons by sediments and soils.* Environmental Science and Technology 16(2): 93-98.
47. Mineat, R. A., R. O. Ball, and R. Church. 1981. *Data base for influent heavy metals in publicly owned treatment works.* EPA-6602-81-220. U.S. Environmental Protection Agency, Cincinnati, Ohio.
48. Newton, D. W., R. Ellis, Jr., and G. M. Paulson. 1976. *Effect of pH and complex formation of mercury (II) adsorption on bentonite.* Journal of Environmental Quality 5: 251-254.
49. Sommers, L. E. 1977. *Chemical composition of sewage sludge and analysis of their potential as fertilizers.* Journal of Environmental Quality 6: 225-232.
50. Spencer, W. F., and M. M. Cliath. 1975. *Vaporization of chemicals.* In R. Haque and V. H. Freed [editors] *Environmental Dynamics of Pesticides.* Plenum Press, New York, New York. pp. 61-78.
51. Spencer, W. F., W. J. Farmer, and W. A. Jury. 1982. *Review: Behavior of organic chemicals at soil, air, water interphases as related to predicting the transport and volatilization of organic pollutants.* Environmental Toxicology Chemicals 1: 17-26.
52. Tabak, H. H., S. A. Quave, C. I. Mashni, and E. F. Barth. 1981. *Biodegradability studies with organic priority pollutant-compounds.* Journal of Water Pollution Control Federation 53(10): 1,503-1,518.
53. U.S. Environmental Protection Agency (U.S. EPA). 1987. *Estimation of the 98th percentile of POTW mean pollutant concentrations in sewage sludge.* U.S. Environmental Protection Agency, Washington, D.C.
54. Williams, D. E., J. Vlamis, A. H. Pukite, and J. E. Corey. 1987. *Metal movement in sludge-amended soils: A nine-year study.* Soil Science 143: 124-131.

9

Farming by Soil

W. E. Larson and P. C. Robert

Because landscapes and soils vary greatly, land use and management must be tailored to fit their specific properties. Soil spatial variability within fields has been widely demonstrated by soil testing results and crop yield differences. There are numerous reasons for the variation in soil characteristics and potentials, including soil forming factors (parent material, topography, vegetation, climate, and time), farming practices (tillage, crop rotations, and fertilization), and erosion (Figure 1). Because the irregular soilscape units do not match well with the symmetrically shaped legal boundaries, several soilscape units usually are managed in the same way. Modern agricultural machinery is best suited for use on rectangular fields.

Field Variability

Field variability is due to natural differences in soil properties and past management. Consider the soil map of a 21.8-hectare (54-acre) field in Jackson County, Minnesota (Figure 2). This field is 1.7 kilometers by 0.13 kilometer. The soils and landscape are characteristic of the glaciated agricultural area of southwestern Minnesota. The drainage class, texture, organic matter content, slope, taxonomic classification, and area of the seven mapping units are listed in table 1. Note that surface textures range from loam to clay loam, organic matter varies from 3.7 to 6.9 percent, and drainage class ranges from well-drained to poorly drained.

Crop yields vary greatly in uniformly managed fields because of soil and landscape differences (Figure 3). In figure 3, corn yields are plotted on an aerial basis for a research field in Redwood County, Minnesota, near the Southwest Branch Experiment Station. In treatment 1, fertilizer was uniformly applied; in treatment 2, no fertilizer was applied; in treatment 3, fertilizer was applied to each soilscape unit, depending on soil type, and herbicide was applied uniformly; and in treatment 4, both fertilizer and herbicide applications were varied according to soil type. In South Carolina's Coastal Plains, corn (*Zea mays* L.) yields varied from 3.7 to 8.0 megagrams per hectare in 1985, 0.1 to 2.8 megagrams per hectare in 1986, and 1.1 to 5.0 megagrams per hectare in 1988 on an 8-hectare site. Wheat (*Triticum aestivum* L.) yields ranged from 3.5 to 6.2 megagrams per hectare, while grain sorghum [*Sorghum bicolor* (L.) Moench] yields varied from 1.7 to 5.3 megagrams per hectare in 1987 (*3*). In Montana, barley yields (*Hordeum vulgare* L.) varied from 1.4 to 3.6 megagrams per hectare, from 1.6 to 4.3 megagrams per hectare, and from 0.8 to 1.1 megagrams per hectare in three uniformly managed fields. Yields of spring wheat ranged from 1.2 to 1.8 megagrams per hectare in a fourth field (*1*). Phosphorus soil tests varied from 18 (medium) to 40 (high) kilograms per hectare, and soil test potassium varied from 290 (high) to 482 (very high) kilograms per hectare in a Minnesota field that had been uniformly managed for years. Past management may also create field variability. Manure has often been regularly spread on areas closest to the livestock pens or barns and easily accessible fields or portions of fields. Spreading mineral fertilizer uniformly for years may actually have increased the variability by overfertilizing some areas and underfertilizing other areas, resulting in significant soil test differences.

Some landscapes are more suited than others to the arrangement of land into management units. Larson and associates (*4*) discuss two

Figure 1. Soilscape variability in Minnesota

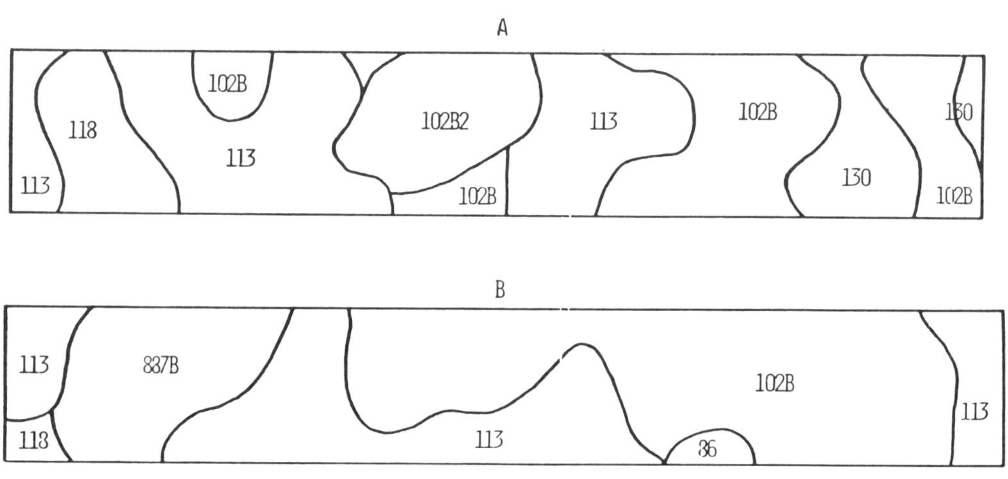

Figure 2. Soil survey map of a field in Jackson County, Minnesota

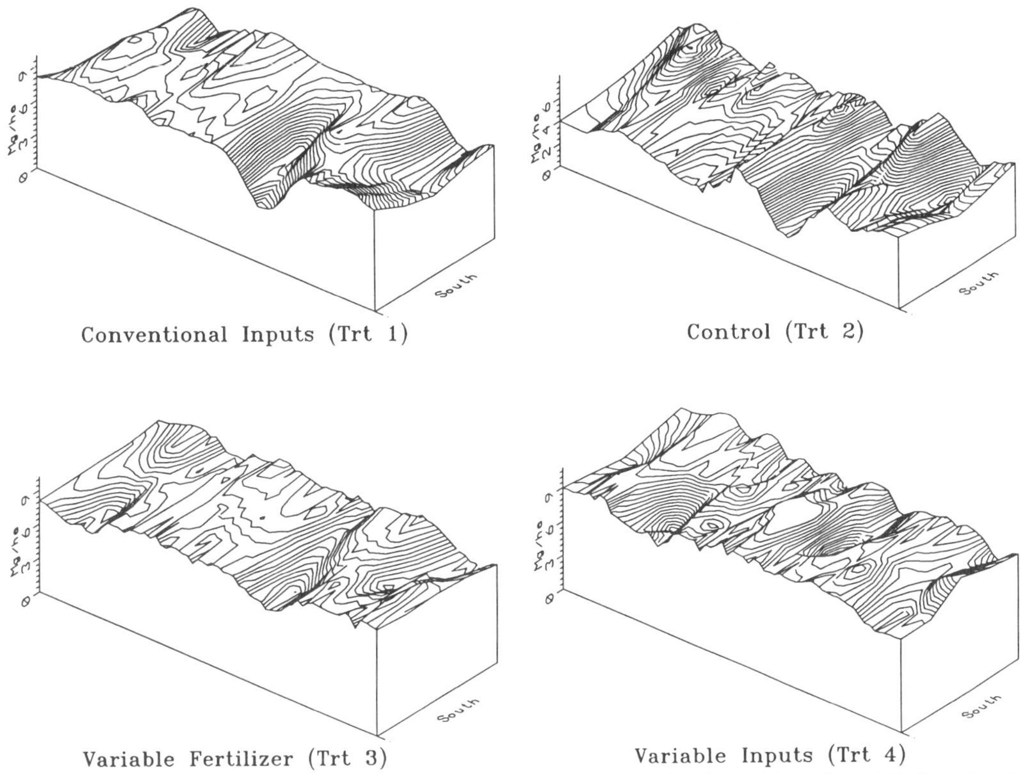

Figure 3. Yield variability at the farming-by-soil research site near the Southwest Experiment Station, Lamberton, Minnesota.

soilscape unit extremes in the north central United States (Figure 4). One has small relief, no major surface outlet, and containment of transported sediment in depressional areas. The soils and slopes may vary in irregular patterns, ranging from deep, medium-textured to shallow coarse-textured soils. Management units usually are arranged in rectangular blocks. The other topographic extreme has distinct slopes and deeply incised stream valleys. In this landscape, erosion may be severe, and a relatively large amount of the sediment may leave the cultivated area and be deposited on a floodplain or carried far away. In these landscapes, management units are often arranged on the contour on similar slopes.

About 12,000 soil series are recognized in the United States. When soil maps are overlaid by landscape maps, the soilscape unit becomes extremely large. It is not feasible to manage each soilscape unit as a separate field with conventional practices. The size of a soilscape unit may vary from 1 hectare to many hectares.

Emerging Technology

The technology is now emerging that will allow workers to operate farm machinery within current field boundaries and to change the management by altering the equipment as it traverses the field (Figure 5). For example, microprocessors using digitized maps in the cab of chemical spreaders can alter fertilizer and herbicide applications at predetermined amounts as the machine moves over the field (Figure 6). The digital maps may be developed from soil surveys, aerial photographs, chemical soil tests, historic crop yields, estimated erosion amounts, and/or other meaningful data. While on-the-go equipment is emerging, quantitative databases and interpretations are not always available.

The terms farming-by-soil, soil-specific

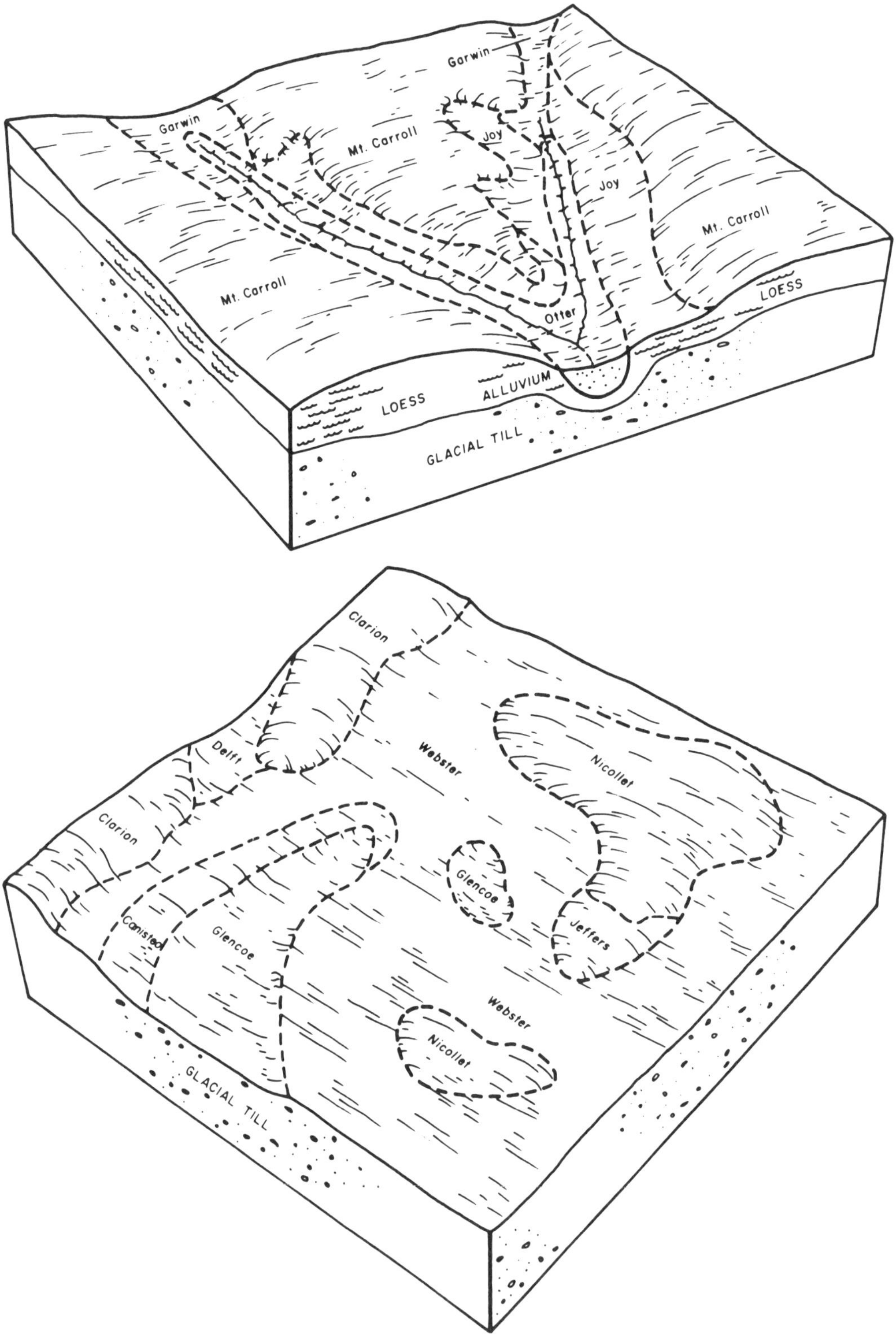

Figure 4. Relation of soils to their underlying materials and landscape in the Webster-Nicollet associations (top) and the Mount Carroll-Otter-Joy association in Minnesota.

Figure 5. Soil-specific chemical spreader.

Figure 6. Microcomputer monitor showing a field-condition map in the cab of a spreader.

farming, precision farming, and farm-by-the-foot have been introduced to describe the soil-specific management concept (*8, 10*). Applying fertilizer according to field variability will optimize soil productivity by selecting grades and rates as a function of soil potential. In addition, it will improve fertilizer recommendations by selecting soil sampling sites more carefully and by representing field-soil fertility variability more precisely. The lack of specificity concerning soil variation and response to the specific needs of soilscape units leads to unwise land use and management and may result in erosion, over- or under-application of chemicals, undue leaching of chemicals, undesirable tillage, and less than optimum production on some soilscape units.

Farming by Soil

The best land use may vary within a field according to soilscape units. For example, in the glaciated areas of the north central United States, gently rolling, fine-loamy Hapludolls with slopes less than 2 percent may be interspersed with coarse-loamy over sandy or sandy skeletal Hapludolls with slopes of 6 to 10 percent. The potential for row crop production of corn and soybeans [*Glycine max* (L.) Merr] on the former may be high, whereas use of the latter soils may be unprofitable for row crop production. Yet, the mixed soilscape units may be farmed in the same way for convenience of machine operation. With a modern, digitized soil survey, it may be possible to arrange management units (fields) to improve land use without losing efficiency in field operation.

Matching management with the soil-specific differences in fields may be an important means of reducing water and environmental contamination with agricultural chemicals. Not only can overapplication of chemicals be better controlled, but areas more sensitive to leaching or surface runoff can be located, and chemical applications can be altered. A software package developed in Minnesota will quickly identify soil mapping units within a field where leaching is a potential hazard. Chemicals could be applied at reduced rates on land strips near streams or water bodies.

Easy-to-use, computer-assisted models using soil, hydrology, and landscape information are needed to best arrange land use and management units. While ownership and legal boundaries may sometimes limit the most desirable arrangement, fitting management into natural ecological units is desirable to maximize natural resource use and conservation. In many cases, two or more adjacent land operators may cooperate in managing the ecological units.

Knowing the nature of the soilscape unit variability within a field is not sufficient without auxiliary databases from which meaningful interpretations and recommendations can be made. One must know how a crop responds to fertilizer on each soilscape unit. The amounts of needed herbicide can often be interpreted using soil characteristics, such as texture, organic matter, and pH. Hydrologic characteristics, along with landscape and soil map characteristics, can be used to interpret water management and erosion control decisions.

Information needed in careful land use and management planning includes soil, landscape, weather, and hydrologic data. Soil information includes water-holding capacity, plant nutrient and chemical characteristics, texture and other physical data, and pH, all by each horizon throughout the solum (*2*). Landscape data include slope length and steepness; landscape shapes, including concavity and convexity; natural waterways; gullies; upslope drainage area; and description of overland flow paths (*5*). Current and historical weather information provides data on the probability of weather events as they affect natural resource management. Required hydrologic data include soil hydraulic conductivities, depth to water table, vadose zone characteristics, and overland flow characteristics. It is essential that all of the data bases are available in such a format that they are compatible for cross-referencing. Development of the geographic data bases and interpretive software is a major research need.

Fertilizer Management

Crop equivalent ratings, estimated corn yields (*11*), and recommended nitrogen fer-

Table 1. Mapping units and description of soils, Jackson County, Minnesota.

No.	Series	Drainage Class	Texture	Slope (%)	Organic Matter (%)	Subgroup	Area (ha)
86	Canisteo	PD	Clay loam	0-2	6.9	Typic Haplaquolls	0.1
102B	Clarion	WD	Loam	2-6	5.0	Typic Hapludolls	9.8
102B2	Clarion	WD	Loam	4-8	4.7	Typic Hapludolls	.14
113	Webster	PD	Clay loam	0-2	7.0	Typic Haplaquolls	6.9
130	Nicollet	MWD	Clay loam	0-2	6.3	Aquic Hapludolls	1.1
887B	Clarion-Swanlake	WD	Loam	2-6	3.7	Entic Hapludoll†	1.5
118	Crippin	SWPD	Clay loam	0-2	6.0	Aquic Haplaquolls	1.0

*PD represents poorly drained, WD represents well drained, MWD represents moderately well drained, and SWPD represents somewhat poorly drained.
†Swanlake soil.

tilizer rates for continuous corn (*7*) for the field described in figure 2 and table 1 are presented in table 2. The crop equivalent rating relates soil properties to crop yields and the inputs necessary to obtain designated yield. The crop equivalent ratings are indexed on a scale of 0 to 100 (*11*). Note that the crop equivalent ratings range from 70 to 100 in the study field. Estimated corn yields range from 6,272 to 9,094 kilograms per hectare. Nitrogen recommendations for continuous corn, using estimated corn yields (*11*) as the yield goal, vary from 112 to 168 kilograms per hectare (Table 2). If the entire field were fertilized for soils with the highest estimated yield (Nicollet, Crippin; 168 kilograms per hectare; 10 percent of area), 31 percent of the area (Webster) would be overfertilized by 22 kilograms per hectare, and 45 percent of the area would be overfertilized by approximately 31 kilograms per hectare (Clarion B). The Canisteo, Clarion-Swanlake, and Clarion B2 areas (12 percent of area) would be overfertilized by 56 kilograms per hectare. In this example, a uniform application of nitrogen fertilizer unavoidably will over- or under-fertilize some soils.

In the case of phosphorus, recommended application amounts range from 10 to 35 kilograms per hectare of phosphate. Phosphorus and potassium recommendations are based on soil tests of samples taken from the field (*7*).

If the entire field is treated with fertilizer and herbicide for the Nicollet and Crippin soils, compared with the differential chemical application for each soil, the savings in fertilizer and herbicide costs would be about \$17.00 per hectare on the 21.8-hectare field. This is similar to previous reports by Robert (*9*). Yields, input costs, and relative net returns in the farming-by-soil research site near the Southwest Experiment Station at Lamberton, Minnesota, are provided in table 3. Field trials conducted in Missouri, Kansas, Montana, Washington, Wisconsin, and Minnesota also show similar or higher savings in fertilizer and herbicide costs (personal communication from Dean Fairchild, research soil scientist, Soil Teq Inc., Waconia, Minnesota).

Herbicide Use

Table 4 presents estimated hydraulic conductivities and adsorption coefficients for two herbicides on two soils.

Table 2. Crop Equivalent Rating (CER), estimated corn yield, and recommended fertilizer application for the test field.

Soil Mapping Unit	CER*	Estimated Corn Yield	Recommended Fertilizer Application† N	P_2O_5	K_2O
		kg/ha			
Canisteo	77	6272	112	10	12
Clarion B	88	7652	134	15	12
Clarion B2	75	7200	112	12	12
Webster	85	8154	134	35	12
Nicollet	100	8781	168	20	12
Clarion-Swanlake	50	4500	45	10	12
Crippin	88	9094	168	20	12

*From Rust and associates (*11*).
†Based on Rehm and associates (*7*).

Sorption of most herbicides on soils is related to the organic matter content, amount and nature of the clay, and pH in the soil. Adsorption of atrazine is related to the pH, clay, and organic matter, whereas adsorption of diuron is most closely related to the organic matter content of the soil. The lower adsorption and higher hydraulic conductivity (5X) combine to significantly increase the risk of herbicide leaching on the Clarion loam. The hydraulic conductivity was estimated from the procedure of Rawls et al. (*6*). Herbicide adsorption was estimated from the soil texture and from the measured pH and organic matter content of the soil. Herbicide application needs to be fine-tuned for each soil to assure effective weed control while minimizing the risk of leaching.

Soil Erosion

Under uniform management, soil mapping units within a field often vary greatly in their susceptibility to erosion and in their vulnerability to damage. Soil characteristics and topographic conditions, as well as management, influence the potential for erosion. Consider the mapping units provided in figure 2 and selected erosion data provided in table 5.

In column 3 of table 5, the estimated annual erosion amounts are given using the universal soil loss equation (USLE) (*12*). The midpoint of the slope percentage range was used to estimate the SL (topographic factor) parameter in the USLE. The assumed C factor (cover and management) was 0.3, which is approximately the value for conventional tillage and corn following corn. Note that erosion varies from 2.9 to 22.4 megagrams per hectare per year. The vulnerability to losses in soil productivity of the soils is presented in column 4 of table 4. It is calculated as 100 multiplied by the slope of the soil productivity versus the soil removed (centimeters) across the 0- to 50-centimeter range (*4*). The Clarion soil has both the greatest (most negative) vulnerability and the greatest erosion rate. Considering the landscape characteristics in this example, it is likely that the sediment eroded from the sloping soils will be deposited at lower elevations on the other cultivated soils.

Using the estimated erosion rates in column 3 of table 4, assuming they continue for 100 years and the V values in column 4, we calculated the percent loss in soil productivity over 100 years and presented them in column 5. Estimated losses in productivity range from 0.4 to 12.6 percent and reflect both the greater erosion losses and greater vulnerability to damage. In the example, the Clarion soil is the most erodible and the most vulnerable to damage.

The need for erosion control practices on the soils with more than a 2 percent slope is ob-

Table 3. Yields, input costs, and relative net returns at the farming-by-soil research site near the Southwest Experiment Station, Lamberton, Minnesota.

	Treatments		
	Conventional	*Farming-by-Soil*	*Check*
Yields (kg/ha)	8,340	9,090	5,080
Input costs ($/ha)	69	76	36
Relative net returns ($/ha)	246	266	155

Table 4. Hydraulic conductivity and adsorption coefficients for two herbicides on two soils.

Soil	*Hydraulic Conductivity* (cm/s)*	*Adsorption Coefficient (K/d)* Atrazine	Diuron
Clarion	2.2×10^{-3}	2	4
Webster	4.2×10^{-5}	9	25

*Estimated from Rawls and associates (*6*).

Table 5. Estimated erosion, vulnerability (V) to damage from erosion, and loss in soil productivity.

Mapping Unit	*Slope (%)*	*Estimated Erosion (Mg/ha/yr)*	*V†*	*Loss in Soil Productivity† over 100 Years† (%)*
Canisteo	0-2	2.9	−.260	0.6
Clarion B	2-6	12.4	−.686	7.0
Clarion B2	4-8	22.4	−.686	12.6
Webster	0-2	3.1	−.290	0.6
Nicollet	0-2	2.9	−.228	0.4
Clarion-Swanlake B	2-6	12.4	−.382§	3.9
Crippin	0-2	2.9	−.226	0.4

*Estimated by the Universal Soil Loss Equation (*12*). A C factor of 0.3 was assumed for all soils, which approximates conventional tillage.

†The vulnerability (V) to loss in soil productivity from erosion. V is 100 times the slope of the soil productivity versus soil (cm) removed curve (*4*).

‡Assumes erosion will continue for 100 years at amount specified in Column 4.

§Calculated on 60% Clarion, 40% Swanlake soils.

vious. The capability to change conservation practices in straight-line farming as tillage machines move from one soil to another is difficult, but not impossible. Tillage machines that alter the roughness or residue cover might be designed. For example, the twist on the shanks of chisel plows might be rotated, or the depth of tillage could be altered to meet the roughness or residue cover needs of each soil condition. Research is needed to find innovative conservation practices applicable for the concept of farming by soil.

A Management Perspective

The examples of fertilizer needs, herbicide sorption, hydraulic conductivity, and erosion for soils within a field in Jackson County, Minnesota, all strongly emphasize that management for crop production needs to be performed on a soil-by-soil basis.

Farming by soil can increase farm profitability, reduce the potential for chemical leaching, and reduce the potential for erosion and sedimentation on parts of the landscape. Modern soil surveys or grid systems delineating meaningful characteristics, microcomputers, and variable application equipment make farming by soil feasible.

Farmers tend to fertilize for the better soils in a field. In the Jackson County example, fertilizing for the Nicollet and Crippin clay loams would result in an excess of approximately 31 kilograms per hectare on the Clarion B and Webster soils and an excess of 56 kilograms per hectare of nitrogen on the Canisteo and Clarion soils. The farming-by-soil management had a $20 per hectare benefit over conventional management at the Southwest Experiment Station research field (Table 3). Expenses included custom service charges, drying costs, and input costs that vary among treatments. The relative net return of the farming-by-soil treatment was weighted by the proportional acreage of each soil condition. The loam soils have a lower adsorption coefficient for atrazine and diuron herbicides. Thus, a herbicide application designed for the clay loam soils may result in excess application for the loam soils. The loam soils have an estimated hydraulic conductivity approximately five times greater than the clay loam soils. Thus, the soils with the greatest likely excess of nitrogen and herbicide also have the greatest potential for leaching.

Because of topographic conditions, the erosion hazard is greatest on the Clarion and Clarion-Swanlake soils, which are also most vulnerable to loss in productivity. Tillage and other erosion control practices need to be designed for the sensitive soils.

Development of farming-by-soil technology is one of the greatest breakthroughs in soil management in a decade or more. While the basic technology is available, it must be fine-tuned and applied. There is a continued need for better soil maps digitized for rapid recall; for better soilscape unit descriptions; and for improved, low-cost application equipment. There is also a need for improvement and development of data bases built on the basic natural resource unit—the soil mapping unit. When the information is applied to soilscape units as they occur in the field, the current concerns over cost of production and environment will be eased considerably.

REFERENCES CITED

1. Carr, P. M., G. R. Carlson, J. S. Jacobsen, G. A. Nielsen, and E. O. Skogley. 1990. *Farming soils, not fields: a strategy for increasing fertilizer profitability.* Submitted to Journal Prod. Agriculture (in review).
2. Grigal, D. F., and R. S. Turner. 1988. *Evaluating impacts of pollutants from the atmosphere.* In *Proceedings, International Interactive Workshop on Soil Resources: Their Inventory, Analysis and Interpretation for Use in the 1990s.* University of Minnesota, St. Paul. pp. 166-191.
3. Karlen, D. L., E. J. Sadler, and W. J. Busscher. 1990. *Crop yield variation associated with Coastal Plain soil map units.* Soil Science Society of America Journal 54: 859-865.
4. Larson, W. E., F. J. Pierce, and R. H. Dowdy. 1983. *The threat of soil erosion to long-term crop production.* Science 219: 458-465.
5. Larson, W. E., and G. R. Foster. 1988. *Erosion and sediment deposition.* In *Proceedings, International Interactive Workshop on Soil Resources: Their Inventory, Analysis and Interpretation of Use in the 1990s.* pp. 222-230. University of Minnesota, St. Paul.
6. Rawls, W. J., D. L. Brakenseik, and B. Soni. 1983. *Agricultural management effects on soil water processes Part I: Soil water retention and Green and Ampt infiltration parameters.* Transactions, American Society of Agricultural Engineers 26: 1,747-1,757.
7. Rehm, G. W., C. J. Rosen, J. F. Moncrief, W. E. Fenster, and J. Grava. 1986. *Guide to computer pro-*

grammed soil test recommendations for field crops in Minnesota. AG-BU-0519, Minnesota Extension Service, St. Paul.

8. Reichenberger, L., and J. Russnogle. 1989. *Farm by the foot.* Farm Journal (mid-March): 11-15.
9. Robert, P. C. 1989. *Land evaluation at farm level using soil survey information systems.* In J. Bouma and A. K. Bregt [editors] *Land Qualities in Space and Time.* Purdoc, Wageningen, The Netherlands. 352 pp.
10. Robert, P. C., and J. L. Anderson. 1986. *Use of computerized soil survey reports in county extension offices.* In A. B. Bottcher and F. S. Zazueta [editors] *Proceedings, International Conference on Computers.* Cooperative Extension Service, University of Florida, Gainsville.
11. Rust, R. H., L. D. Hanson, and J. L. Anderson. 1984. *Productivity factors and crop equivalent ratings for soils of Minnesota.* AG-BU-2199. Minnesota Extension Service, St. Paul. p. 57.
12. Wischmeier, W. H., and D. D. Smith. 1978. *Predicting rainfall erosion losses—a guide to conservation planning.* Agriculture Handbook No. 537. U.S. Department of Agriculture, Washington, D.C.

10

Energy Efficiency and Sustainability of Farming Systems

David Pimentel and G. H. Heichel

Agricultural ecosystems, like natural ecosystems, are networks of energy and material flows. Green plants capture solar energy and convert it into chemical energy for use as food by humans and other consumers. An adequate and sustainable supply of food is basic to the survival and progress of human civilization.

Sustainability of food, soil, water, or other natural resources can be described by physical, biological, and economic principles; a unifying definition is elusive. Rather than debate definitions, we propose the concept that a sustainable resource is one that has the opportunity to persist and can be reused or renewed without negatively affecting the environment. By analogy, sustainable systems for producing food and fuels are persistent and renewable without detrimental environmental effects.

For about one million years humans lived in hunter-gatherer systems. Most hunter-gatherer systems were totally sustainable because human populations were relatively stable and never surpassed more than about 6 million people. During this period, humans were completely dependent on solar energy. Only during the last 10,000 years have humans harnessed draft animals, water, wind, and fossil fuel power to augment human energy. Fossil energy was first used in agriculture about 300 years ago.

One of the major forces that moved humans from hunter-gatherer to slash-burn agricultural systems involved world population growth (*1*) (Figure 1). As the population increased, a higher rate of food production per hectare was necessary than was possible with hunter-gatherer systems. Early slash-burn agriculture, with a 20-year rotation, was sustainable. Solar energy and the natural ecosystem over a 20-year fallow period restored the productivity of a piece of land used for food production. The land could be used for about 2 years before being returned to fallow. This sustainable slash-burn agricultural system required about 10 hectares of land to provide food from about 1 hectare of food crops.

As the world population grew, humans abandoned slash-burn agriculture and moved to more intensive agriculture, resulting in the introduction and increased use of fossil energy (Figure 2). Fossil energy was substituted for cropland, animal power, and human labor (*10*). Farmers now produce 30 to 40 times as much food and feed on 1 hectare than was possible with the 10-hectare slash-burn system in which 1 hectare was devoted to crops. Although contemporary, energy-intensive agricultural systems are highly productive, their sustainability can be questioned for at least three reasons. Because fossil energy supplies are not renewable, current intensive agriculture is not sustainable (*32*). In addition, serious soil erosion in contemporary agricultural systems further limits their sustainability (*28, 33*). The recent farm income crisis, the instability of the rural economy, and the high costs of commodity support programs prompts questions about the economic stability of contemporary agricultural systems.

Rapid growth in the world population exacerbates the severity of the energy and soil resource problem (Figure 1 and 2). The world's human population has reached about 5.2 billion (*35*) and is projected to exceed 6.2 billion by the turn of the century (*8*). It may grow to 12 billion by 2100 and probably will not stabilize until it reaches 15 to 20 billion (Figure 2).

More than 1 billion humans are now malnourished because of world food shortages (*23*) and concurrent problems of food distribution. Shortages of land, water, fossil energy, and biological resources already plague many parts of the world (*28, 30*). Fossil energy resources per capita in the world will slowly decline soon after the year 2000 (Figure 2). If the earth is to support humans with a high standard of living, like that in the United States, and using solar energy, then the world population would have to be much less than the present 5.2 billion. Careful planning and about 200 years will be needed to achieve a stable population with a high standard of living (*28*). The need for technologies and policies to achieve stable populations and environmentally sound, sustainable agricultural production systems is urgent.

Our objective is to compare scenarios of energy flow in food production systems throughout agricultural development. The evo-

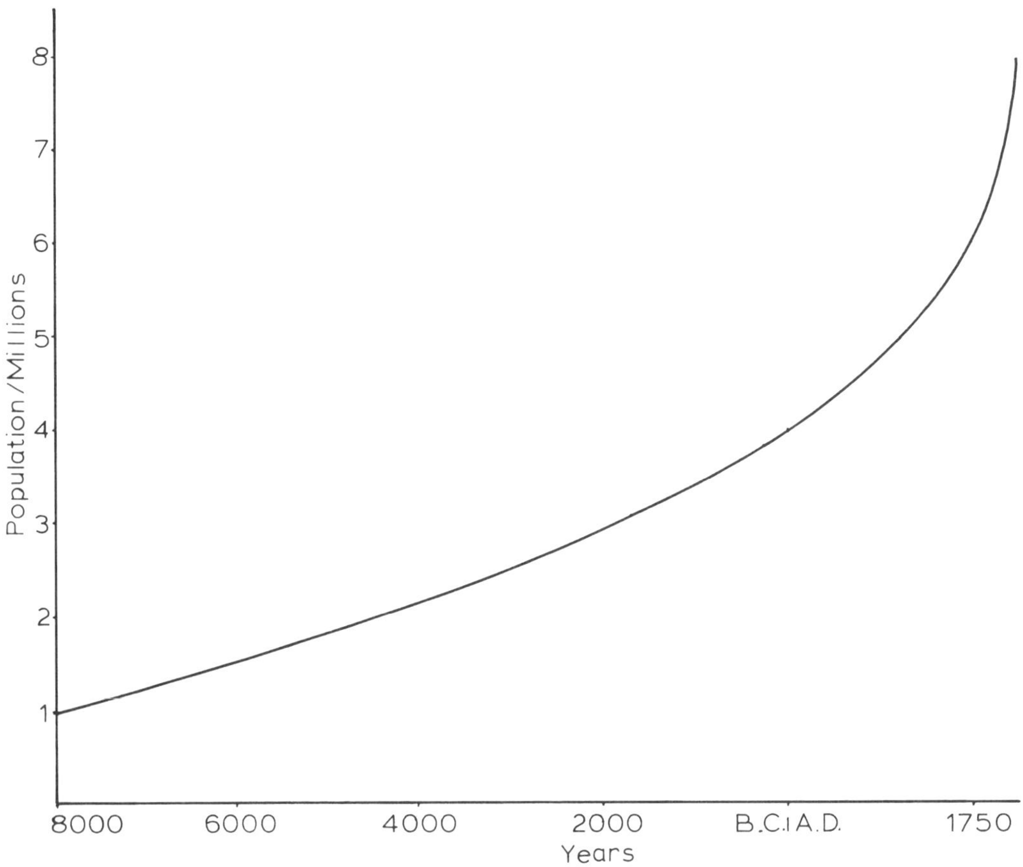

Figure 1. World population growth after the introduction of agriculture about 10,000 years ago. The population numbered only about 1 million at the discovery of agriculture (*2, 4*).

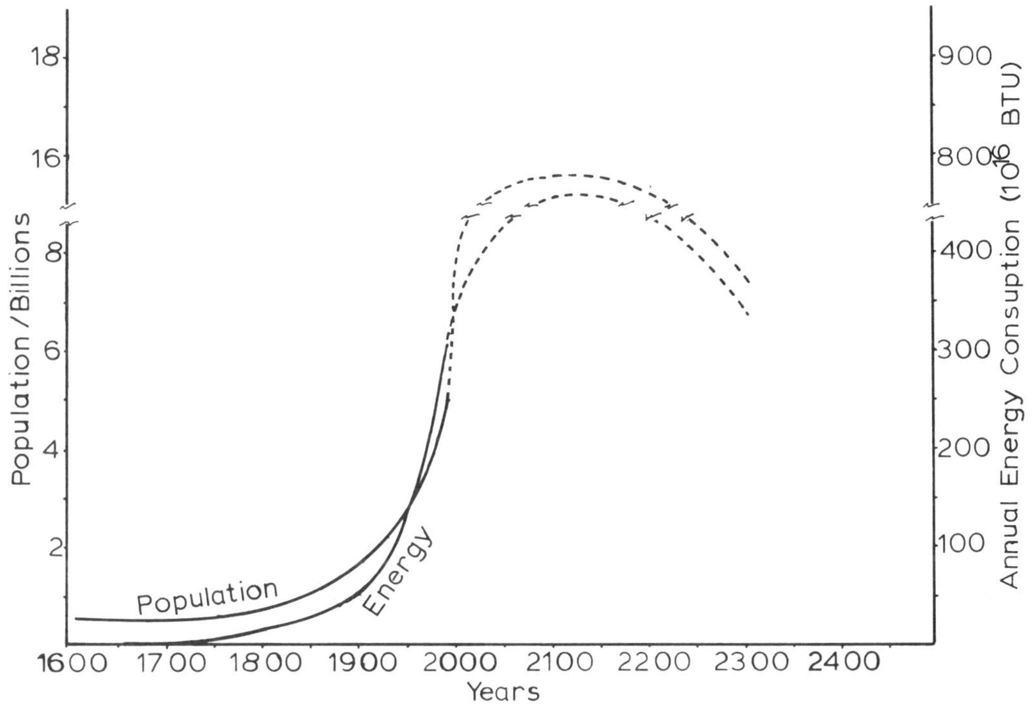

Figure 2. Rapid human population growth associated with fossil energy use. Estimated world population (—) from 1600 to 1987 and projected numbers (----) to the year 2300. Estimated fossil energy consumption (—) from 1650 to 1987 and projected (----) to the year 2300.

lution from sustainable systems of low productivity to highly productive systems of dubious sustainability will be shown. Two technologies will be examined that might foster the development of agricultural production systems that are more sustainable for energy use than those now generally prevalent. Finally, opportunities for crafting public policies to facilitate the adoption of more sustainable agricultural systems will be discussed.

Soil Erosion and Loss of Agricultural Sustainability

At a time when the world population is growing rapidly and the need for more land to produce food is escalating, valuable land is being degraded by erosion (*31*). Soil erosion is of great concern because soil reformation is extremely slow. About 500 years (with a range of 200 to 1,000 years) are required to renew 2.5 centimeters of soil—a renewal rate of about 1 metric ton of topsoil per hectare per year (*5, 18, 20, 21*). Some speculate that the United States has lost approximately 8 centimeters of topsoil since farming began (*9*).

Erosion rates vary from region to region because of differences in topography, rainfall, wind, and type of agricultural practices used. In China, for example, the average annual soil loss is about 40 tons per hectare per year, while the United States average is 18 tons per hectare per year (*31, 42*). Worldwide degradation of land by erosion and other factors causes the irretrievable loss of an estimated 6 million hectares annually (*41*).

Although reduced soil depth is a serious concern because it is cumulative, other factors associated with erosion are equally serious in reducing productivity of the land. These other factors include rapid water runoff, loss of water-holding capacity, organic matter, soil nutrients, and soil biota (*31*). Soil degradation worldwide is being offset by substituting fertilizers, irrigation, pesticides, and other fossil fuel inputs for the natural endowments of topsoil. The substitution of manufactured and purchased inputs for a productive soil resource increases the flow of fossil energy through the agricultural system.

Soil erosion is rated by many as the most serious environmental problem in the world (*41*). Erosion and the abandonment of agricultural land is a prime cause of world deforestation (*34*), although other socioeconomic and political factors contribute as well.

Energy Flow in Hand-Powered, Sustainable Agricultural Systems

The foundation of agricultural production systems rests on the unique capability of plants to convert solar energy into stored chemical energy. The success of agricultural production is measured by the amount of solar energy that is captured and converted into food per unit land area as a result of manipulating plant, land, and water resources using human, animal, and fossil energy power.

The incident solar energy radiated during the year in temperate North America averages about 14 billion kilocalories per hectare (*37*). During a 4-month summer growing season in the temperate region, nearly 7 billion kilocalories reach an agricultural hectare. Under favorable moisture and soil nutrient conditions, corn (*Zea mays* L.) is one of the most productive food and feed crops. For example, corn can produce about 2,000 kilograms per hectare under a slash-burn system (*22*) or upward of 7,000 kilograms per hectare in an intensive system. An equal amount of biomass as stover is produced in both systems. Converted to heat energy, the slash-burn and intensive systems produce 18 million and 63 million kilocalories, respectively. This represents from 0.1 to 0.5 percent of the incident solar energy annually.

Corn produced by hand employing slash-burn agricultural technology, constitutes a sustainable agricultural system if it is practiced with an environmentally sound technique, such as a period of fallow. The only fossil energy input is used to produce the axe and hoe; however, these tools could be produced using charcoal, and the system could thus depend totally on solar energy. A total of 1,144 hours of human labor is required to produce the 1,944 kilograms per hectare of corn in this system (*24, 29*). The only other input in this system is 10.4 kilograms of corn seed. We assume that the 1,144 hours of labor represent nearly 60 percent of the total labor output for an adult per year. The farmer is assumed to consume about 3,000 kilocalories per day of food and requires about 6,000 kilocalories of fuelwood for cooking and preparing food. No other input for the farmer is charged to this system. The energy output/energy input for this system is 12.9 (Table 1), using energy budget methods previously described (*24, 29*).

There is a need, however, to include in the energy budget the solar energy radiated to the 10 hectares of land that was fallowed to regain its productivity after use in producing corn on 1 hectare for 2 years. For this system, we assume that, on average, 0.1 percent of the photosynthetically active solar energy is captured annually by the plants growing on the fallow hectares. This energy, then, is charged to this sustainable agricultural system. Therefore, the total energy input, assuming the 140 million kilocalories of solar energy captured on the 10 hectares of land, is calculated to be 140.8 million kilocalories. The output/input ratio changes to 0.08:1 when the solar input in the total land area is included (*29*). However, it must be emphasized that this system is sustainable and requires no fossil energy input, but clearly this agricultural system is land-extensive.

Energy Flow in a Draft Animal Agricultural System

If some of the 1,144 hours of labor in the hand-powered corn system were replaced with nearly 200 hours of ox power, then the human labor input might be reduced to 380 hours (Table 2). Hence, the labor input of 201,000 kilocalories is a large input in the system. The feed needed to supply the ox for about 200 hours worth of work, with a total work capacity of 1,600 hours per year, was 150 kilograms of concentrate (corn) and 300 kilograms of forage (*27*).

The total energy for the ox system is 1.7 million kilocalories with an assumed corn yield of 1,944 kilograms per hectare (Table 2). Without a legume in the rotation system and without dung from the ox, a continuous corn system would yield only 941 kilograms per hectare (*24*). However, the data in table 2 indicate an output/input ratio of 6.3:1. The ratio is about half of the value for the hand-produced corn (Tables 1 and 2).

The 150 kilograms of concentrate required by the ox is derived from the 1,944 kilograms corn yield and, thus, reduces the net yield. The ox consumes forage from 2 hectares of pasture on marginal land. Perhaps 20 percent of the dung produced by the ox could be applied to the hectare of corn without depleting the nutrients and productivity of the pasture. Human wastes are also assumed to be applied to the land.

The corn is assumed to be grown in rotation after a legume green-manure crop with agronomic characteristics of sweet clover (*Melilotus alba*) or vetch (*Vicia villosa*). The legume will provide the minimum nitrogen needs (60 kilograms per hectare) of the corn (*38*) and will also aid in controlling soil erosion and adding organic matter to the soil (*33*).

Thus, 4 hectares is the minimum amount of land area needed to keep this system sustainable. This is lower than the 10 hectares required for the slash-burn system, but it is still land extensive. When the energy input in the 4 hectares is added to the input in the hectare of corn, the total energy input is estimated to be 43.7 million kilocalories. Thus, the output/input ratio is 0.24:1. This value is three times greater than the slash-burn system (Tables 1 and 2).

Table 1. Material and energy inputs and outputs per hectare for corn production in Mexico using only human labor.*

Production Factor	*Quantity*	*kcal*
Input		
Labor	1,444 h	766,500‡
Axe and hoe	16,570 kcal†	16,570
Seeds	10.4 kg†	46,800
Total		829,870
Output		
Yield (grain dry matter)	1,944	10,683,820
kcal output/kcal input		12.87

*See Lewis (*24*), and Pimentel and Pimentel (*33*) for additional explanation of calculations.
†Estimated.
‡See text for assumptions for calculating kcal input.

Table 2. Material and energy inputs and outputs per hectare for corn production in Mexico using oxen.*

Production Factor	*Quantity*	*kcal*
Input		
Labor	383 h	201,480‡
Ox	198 h	
Concentrate	150 kg	525,000‡
Hay	295 kg	885,000‡
Machinery	41,400 kcal†	41,400
Seeds	10.4 kg†	46,800
Total		1,699,680
Output		
Yield (grain dry matter)	1,944 kg	10,683,820
kcal output/kcal input		6.3

*See Lewis (*24*), and Pimentel and Pimentel (*33*) for additional explanation of calculations.
†Estimated.
‡See text for assumptions used in calculating these inputs.
§$0.12/kg ($3/bu).

Energy Flow in a Draft Animal Agroforestry System

This agroforestry system is similar to the draft animal agricultural system in terms of labor, ox power, machinery, and seeds (Table 2). Using the agroforestry system, however, 0.5 hectare is planted to corn and the other 0.5 hectare to a leguminous tree, Leucaena (*19, 40*). The planting design includes two rows of corn alternated with two rows of trees on the contour. The corn in this agroforestry system, which is planted at twice the plant population used in the draft animal agricultural system, may lead to unfavorable interspecies competition (*22*).

Similar to the previous ox-powered system, the forage for the ox is provided by 2 hectares of marginal land. Again, the concentrate comes from the corn grain. To help maintain phosphorous and potassium fertility, about 20 percent of the dung from the ox is applied to the corn. The leguminous tree roots may also supply some phosphorus and potassium from deep in the soil (*35*). Human wastes are also assumed to be applied to the land.

The corn in this agroforestry system is planted at twice the plant population of the previous draft animal system; thus, the same number of corn plants are present in this agroforestry system as in the previous ox-powered system. We are also assuming a similar yield of 1,944 kilograms per hectare (Table 2).

Competition between the Leucaena and corn is reduced at planting by cutting the tree back to a stump measuring about 8 centimeters before the corn is planted. The tree produces 4,500 kilograms per hectare of biomass (*36*). Of this total biomass, about 2,500 kilograms of leaves and small twig mulch are worked into the soil for biological nitrogen and organic matter for soil and water conservation. The remaining 2,000 kilograms of Leucaena are harvested as stems for fuelwood. This system provides some of the fuelwood needs of the farmer's family, an advantage when compared with the previous ox-powered system.

Planting the Leucaena on the contour, plus mulching with 2,500 kilograms of leaves and twigs, should limit erosion to only 1 ton per hectare per year (*19*). About two-thirds of the nitrogen in the Leucaena biomass is contained in the 2,500 kilograms of leaves and twigs (*36*). This results in about 60 kilograms per hectare applied to the land, which is similar to the amount applied as a result of the previous ox-powered system.

The 150 kilograms of concentrate required by the ox are derived from the 1,944 kilograms of corn grain and, thus, reduces the net yield. The ox consumes forage from 2 hectares of pasture on marginal land. The total energy for this ox system is calculated to be 1.7 million kilocalories, with an assumed 1,944 kilograms per hectare of corn (Table 2). The data indicate an output/input ratio of 6.3, which is the same as the previous system (Table 2).

The total land area needed to keep this system sustainable is 3 hectares. This is lower than the 4 hectares for the previous ox-powered system, but is still relatively land-extensive. When the energy input in the 3 hectares is added to the input in the hectare of corn, the total energy input is estimated to be 29.7 million kilocalories. Thus, the output/input ratio is 0.36. This value is 4.5 times greater than the slash-burn system (Tables 1 and 2).

Energy Flow in Contemporary Intensive Agriculture

The energy flow in tractor-powered agriculture is distinctly different from that of the hand- and ox-powered agricultural systems (Tables 1-3). Typically, U.S. corn production is heavily mechanized. The human labor input is dramatically reduced to only 10 hours compared with 1,144 hours for the hand-powered system and 380 hours for the ox-powered system (Tables 1-3).

Balanced against this low labor input is the significant increase in fossil energy input needed to run the machines that replace labor and land. In 1988, the energy inputs (mostly fossil fuel) required to produce 1 hectare of corn in the United States averaged about 7.8 million kilocalories, or the equivalent of about 785 liters of oil (Table 3). Based on a corn yield of about 6,500 kilograms per hectare, or the equivalent of 26 million kilocalories of energy, the output/input ratio is about 3.3:1 (*29*). If the amount of solar energy and land area were included in the calculation of the biomass produced and eventually converted into soil and coal, the ratio would be at least three orders of magnitude less than the 3.3:1 reported.

Even if we forego an economic appraisal, the contemporary U.S. corn production system is of questionable sustainability compared with the less technologically developed systems briefly illustrated here and discussed elsewhere (*10*). This is due to the system's dependence on nonrenewable fossil fuels and its susceptibility to high soil erosion rates.

Technological alternatives are available to foster more sustainable crop productions systems than the example given for corn. Space does not allow an exhaustive discussion of possible alternatives, but two will be used as examples because they are currently being covered by the media and in public policy forums. The first is ethanol production from corn grain, which is proposed to enhance the demand for surplus corn production. The other is the use of legume-based crop rotations, which are featured in proposals to modify commodity support programs.

Ethanol Production from Corn Grain

Production of ethanol from agricultural crops has been proposed as a technology that will make agriculture self-sustaining for fuel energy (*6, 7, 25, 26*). The conversion of sugars and starches to ethanol by fermentation is an established technology (*12*).

Corn production requires significant economic and fossil energy inputs (*29*). This is important because the corn biomass constitutes one of the major energy and economic costs in producing ethanol. For example, producing 1,000 liters of ethanol requires 2,700 kilograms of corn (Table 4), which requires an average of slightly less than 0.5 hectare of land (Table 3).

Once the corn is harvested, three sources of energy costs are involved in its conversion to ethanol: transport to the plant, the conversion itself, and pollution control. These costs, in both energy and economic terms, are given in table 4 for a modern, 200-million-liter-per-year chemical plant. The largest energy inputs are for corn production (3.3 million kilocalories) and for the coal (4.6 million kilocalories) used in the fermentation-distillation process (Table 4). The total energy input to produce 1,000 liters of ethanol is 8.6 million kilocalories, but the ethanol has an energy value of only 5.1 million kilocalories. Thus, a net loss of more than 3.0 million kilocalories occurs for each 1,000 liters of ethanol produced. The energy credit given for the distiller's grain by-product would reduce the net loss by about 700,000 kilocalories, but only if the distiller's grain is fed to ruminant livestock (*7*). Fermentation of crop residue to produce ethanol is an even less viable technology (*12*).

Clearly, producing ethanol is not self-sustaining and, in fact, requires 167 percent more energy than is produced (Table 4). Also, the costs of production per liter are about two-fold more than those of gasoline at 1989 crude oil prices. No ethanol would be produced with current technologies if there were no programs to subsidize it.

Table 3. Material, energy, and economic inputs per hectare for conventional corn production in the United States.

Production Factor	*Quantity*	*kcal (1000)*	*Cost ($)*
Input			
Labor (h)	10	7	50.00
Machinery (kg)	55	1,485	91.00
Fuel (L)	115	1,255	38.00
N (kg)	152	3,192	81.00
P (kg)	75	473	53.00
K (kg)	96	240	26.00
Lime stone (kg)	426	134	64.00
Corn seeds (kg)	21	520	45.00
Insecticides (kg)	1.5	150	15.00
Herbicides (kg)	2	200	20.00
Electricity (10^3 kcal)	100	100	8.00
Transport (kg)	322	89	32.00
Total		7,845	523.00
Output			
Yield (grain dry matter)	6,500 kg	26,000	
kcal output/kcal input		3.31	

*See Pimentel and associates (*34*) for additional explanation of calculations.

Table 4. Material, energy, and economic inputs per 1000 liters of ethanol made from U.S. corn grain.

Production Factor	*kg*	*kcal (1000)*	*Cost ($)*
Input			
Corn	2,700	3,259	217§
Transport of Corn	2,700	325†	32
Water	160,000‡	90‡	20
Stainless Steel	6‡	89‡	10
Steel	12‡	139‡	10
Cement	32‡	60‡	10
Coal	660‡	4,617‡	40
Pollution Costs	—	—	60
Total		8,579	399
Output			
Ethanol (1000 liters)	789	5,130	

*See Pimentel and associates (*34*) for additional explanation of calculations.
†Transport estimated to be 10% of production energy.
‡From Slesser and Lewis (*39*).
§$0.12/kg ($3/bu).

Legume-Based Crop Rotations

Legumes are among the most fuel-efficient crops grown in U.S. agriculture (*13*). Compared with contemporary corn production (Table 3), crops like alfalfa (*Medicago sativa* L.) (Table 5) require less than one-half the fossil energy resources. The energy savings are attributable to several factors. Chief among these is the ability of well-nodulated legumes to symbiotically fix 40 to 90 percent of their nitrogen needs from the atmosphere (*15*), thus freeing them from reliance on manufactured fertilizer for nitrogen nutrition. In addition, a different machinery mix and set of field operations results in many legumes, requiring both less liquid fuel and less energy embodied in equipment than corn (Tables 3 and 5). When forage legumes, like alfalfa and the clovers, are properly managed, they not only are largely sustainable for their own nitrogen requirements but also can provide substantial nitrogen to subsequent crops grown in rotation (*15*). Thus, the use of legumes in rotation with such crops as corn and wheat (*Triticum aestivum* L.) that are normally fertilized with manufactured formulations of nitrogen offer the possibility of reducing the requirements for fossil fuels and fostering the adoption of crop production practices more sustainable than those characteristic of monoculture corn, as the following examples illustrate (*11*).

Continuous corn culture is a common but relatively energy-intensive management technique in Minnesota and elsewhere in the Midwest. Continuous corn is annually subsidized with nitrogen fertilizer produced at the expense of nonrenewable energy. An energy budget for corn can be computed by procedures described previously (*10*). Using this procedure, the fossil energy use of continuous corn is about 7.0 million calories per hectare daily (rotation 1, Table 6), and 50 percent of the energy use is attributable to nitrogen fertilizer (*11*). The contrast in energy usage between continuous corn in table 3 and that in table 6 is attributable to the former being averaged over all U.S. corn-producing areas, whereas the latter represents values typical of Minnesota.

Because soybeans [*Glycine max* (L.) Merr.] usually require no nitrogen fertilizer, alternating 1 year of soybeans with 2 years of corn reduces fossil energy use by 26 percent (*11*), a substantial savings in energy (rotation 2, Table 6). In contrast to perennial forage legumes, however, soybeans do not leave nitrogen in the soil for succeeding crops (*15*).

Substituting a forage legume for soybeans in the corn rotation greatly reduces nitrogen fertilizer needs. The protein in the roots and unharvested regrowth of a 2- or 3-year-old alfalfa stand is equivalent to 210 kilograms of nitrogen per hectare. About 170 kilograms per hectare represents nitrogen fixed by symbiosis, using no fossil energy; the remaining 40 kilograms per hectare are derived from soil organic matter.

Because part of the nitrogen fixed by symbiosis is a net accretion beyond that needed for alfalfa growth, it becomes an energy credit for allocation in a subsequent crop. With rota-

Table 5. Material and energy use per hectare in baled alfalfa hay production in southeastern Minnesota*.

Production Factor	*Quantity*	*Kcal*	*Percent of Total*
Input			
Labor	7.2 h		
Machinery	852 kg	19,906	.7
Liquid fuel	92 L	1,050,090	35.1
Phosphorus (P_2O_5)	34 kg	37,400	1.3
Potassium (K_2O)	68 kg	108,800	3.6
Limestone	2,652 kg	835,390	27.9
Transportation	3,655 kg	939,335	31.4
Total		2,990,921	100.0
Output			
Dry matter yield (15% H_2O)	10,000 kg		
Energy output/energy input		7.2	

*See Heichel (*14*) for additional explanation.

Table 6. Intensity and efficiency of energy use in continuous and rotational cropping systems in Minnesota incorporating annual grain and perennial forage legumes*.

	Rotation			
	1	*2*	*3*	*4*
	Continuous Corn	*2 Corn-Soybean*	*3 Corn-2 Corn-Oats-2 Alfalfa*	*3-Soybean Wheat-3 Alfalfa*
Fossil energy use (Mcal/ha-day)	7.0	5.2	4.3	3.9
Crop yield (kg dry matter/ha)	8,721	6,979	9,010	7,552
Crop energy yield Fossil energy use (kcal/kcal)	6.1	6.7	7.8	8.3

*See Heichel (*11*) for additional explanation.

tion 3 (Table 6), the nitrogen becomes available during the first and second years of corn and may reduce the nitrogen fertilizer requirement of the crop by 50 percent. Compared with that for continuous corn production, the fossil energy use for fertilizer, direct fuel consumption, and other inputs in rotation 3 (Table 6) are reduced nearly 39 percent. Lengthening the rotation to incorporate both an annual and perennial legume achieves greater energy reductions. The fossil energy use of rotation 4 (Table 6) is 44 percent less than that of continuous corn production. The largest decrease in fossil energy use is attributable to a reduced need for nitrogen fertilizer, but fuel and other inputs are also conserved. The ratio of crop energy yield to fossil energy use is greater in rotations than in monocultures, indicating that rotational systems have a more favorable net energy balance and are more sustainable.

One criticism of rotations is that their average annual economic yield over the life of the rotation is often less than that of continuous corn. This need not always be the case, especially for soybean-corn rotations (*3, 17*). Moreover, rotations are becoming increasingly cost-effective because of the demand for forage crops and the beneficial characteristics of rotations that are often economically unvalued. Chief among these beneficial but typically unvalued characteristics is the reduction in soil erosion attributable to the inclusion of a forage legume in rotation or the use of noninversion tillage methods between successive crops.

This discussion has illustrated that the use of legume-based crop rotations would foster more sustainable crop production practices. Other technologies undoubtedly are "on-the-shelf" or nearing technical feasibility. Using crop rotations as an example, the barriers to adoption of cropping systems that are more sustainable (at least for energy use and soil loss) than those in contemporary use can be considered.

Public Policy and Agricultural Sustainability

A major factor limiting the diversification of farming systems to include rotations that incorporate legumes is the current U.S. commodity support program (*16*). Under current (1985 farm bill) commodity program regulations, producers must maintain a base acreage of program-supported crops (e.g., corn, wheat) to qualify for maximum subsidy payments on those crops in the future. The commodity program regulations prevent the use of crop rotations that incorporate legumes in regions where the rotation could be an economic alternative for the farmer or an environmental benefit to society. The commodity program penalizes farmers who choose to adopt rotations that contain nonprogram crops and are adverse to societal interests in sustainable agriculture and environmental quality. The need to couple commodity support and environmental policy is evident. Crop commodity policy directly influences energy use and flow in agricultural production systems and, moreover, the sustainability of natural resources and food production.

In Summary

We have compared energy flows in sustainable, low-yielding, and technologically simple agricultural systems with those of less sustainable, higher yielding, and technologically complex systems. The sustainable systems tend to be extensive in land requirements, low in loss of soil to erosion, and low in fossil energy use. Contemporary corn production in monoculture is also highly productive, but highly reliant on nonsustainable liquid fossil fuel resources. In addition, the rates of soil erosion in contemporary agriculture exceed those of soil formation and are nonsustainable by that criterion. Fermenting grain to ethanol to foster sustainability is not viable because the energy and economic costs of ethanol production exceed the benefits. Crop rotations based on grain and forage legumes as renewable and sustainable source of nitrogen are an available and proven technology that also reduces soil erosion. However, agricultural policy, in the form of commodity support programs, mitigates against the adoption of rotations where they are economically viable or a benefit to society. Improvements in the sustainability of agricultural systems are possible with existing technologies, but such improvements must be fostered by consistent farm and environmental policies.

CONTRIBUTIONS

This chapter is a joint contribution of the New York Agricultural Experiment Station; the Plant Science Research Unit, U.S. Department of Agriculture, Agricultural Research Service; and the Minnesota Agricultural Experiment Station (Science Journal Series No. 17792).

REFERENCES CITED

1. Boserup, E. 1981. *Population and technological change.* University of Chicago Press, Chicago, Illinois. 255 pp.
2. Coale, A. J. 1974. *The history of the human population.* Science of America 231: 40-51.
3. Cramer, C. 1988. *222-bushel corn—without chemicals.* The New Farm 10: 12-15.
4. Deevey, E. S., Jr. 1960. *The human population.* Science of America 203: 195-204.
5. Elwell, H. A. 1985. *An assessment of soil erosion in Zimbabwe.* Zimbabwe Science News 19: 27-31.
6. Energy Research Advisory Board. 1980. *Gasohol.* Report of the Gasohol Study Group. Washington, D.C.
7. Energy Research Advisory Board. 1981. *Biomass energy.* Washington, D.C.
8. Food and Agriculture Organization, United Nations. 1981. *Agriculture: Toward 2000.* New York, New York. 134 pp.
9. Handler, P., editor. 1970. *Biology and the future of man.* Oxford University Press, New York, New York.
10. Heichel, G. H. 1973. *Comparative efficiency of energy use in crop production.* Bulletin 739. Connecticut Agricultural Experiment Station, New Haven. 26 pp.
11. Heichel, G. H. 1978. *Stabilizing agricultural energy needs: Role of forages, rotations, and nitrogen fixation.* Journal of Soil and Water Conservation 33: 279-282.
12. Heichel, G. H. 1979. *American agriculture in an energy perspective.* In *Proceedings. Thirty-fourth Corn and Sorghum Research Conference.* American Seed Trade Association. Chicago, Illinois. pp. 104-118.
13. Heichel, G. H. 1980. *Energy for agriculture.* In L. S. Murphy, E. C. Doll, and L. F. Welsch [editors] *Moving Up the Yield Curve—Advances and Obstacles.* Special Publication. 39. American Society of Agronomy, Madison, Wisconsin. pp. 91-103.
14. Heichel, G. H. 1982. *Energy analyses of forage production systems.* Journal of Animal Science 54: 869-876.
15. Heichel, G. H. 1987. *Legume nitrogen: Symbiotic fixation and recovery by subsequent crops.* In Z. Helsel [editor] *Energy and World Agriculture Handbook, Volume II: Energy in Plant Nutrition and Pest Control.* Elsevier Science Publishers, Amsterdam, The Netherlands. pp. 63-80.
16. Heichel, G. H. 1990. *Legumes as a nitrogen source: Implications for nitrate contamination of groundwater.* In *Agriculture Approaches to Reduce Agrichemical Contamination of Groundwater.* Office of Technology Assessment, Washington, D.C.
17. Helmers, G. A., M. R. Langemeir, and J. Atwood. 1986. *An economic analysis of alternative cropping systems in east-central Nebraska.* American Journal of Alternative Agriculture 4: 153-158.
18. Hudson, N. W. 1981. *Soil conservation.* Cornell University Press, Ithaca, N.Y.
19. Kidd, C. V., and D. Pimentel, editors. 1990. *Integrated resource management: Agroforestry Systems.* Academic Press, San Diego, California.
20. Lal, R. 1984. *Productivity assessment of tropical soils and the effects of erosion.* In F. R. Rijsberman and M. G. Wolman [editors] *Quantification of the Effect of Erosion on Soil Productivity in an International Context.* Delft Hydraulics Laboratory, Delft, The Netherlands. pp. 70-94.
21. Lal, R. 1984. *Soil erosion from tropical arable lands and its control.* Advances in Agronomy 37: 183-248.
22. Lal, R. 1989. *Agroforestry systems and soil surface management of a tropical alfisol: I. Soil moisture and crop yields.* Agroforestry Systems 8: 7-29.
23. Latham, M. C. 1984. *International nutrition and problems and policies.* In M. Drosdoff [editor] *World Food Issues.* New York State College of Agriculture and Life Science, Ithaca, New York. pp. 55-64.
24. Lewis, O. 1951. *Life in a Mexican village: Tepoztlan restudied.* University of Illinois Press, Urbana. 512 pp.
25. Moreira, J. R., and G. E. Serra. 1985. *The Brazilian National Alcohol Program—Incentives and subsidies.* International Institute of Environment and Development, San Diego, California.
26. Moreira, J. R., and J. Goldemberg. 1981. *Alcohol—its use, energy and economics—a Brazilian outlook.* Research Management Optim. 1: 273-279.
27. Morrison, F. B. 1956. *Feeds and feeding.* The Morrison Publishing Company, Ithaca, New York. 1,165 pp.
28. Pimentel, D. 1987. *Technology and natural resources.* In D. J. McLaren and B. J. Skinner [editors] *Resources and World Development.* John Wiley and Sons, New York, New York. pp. 679-695.
29. Pimentel, D., A. F. Warneke, W. S. Teel, K. A. Schwab, N. J. Simcox, D. M. Ebert, K. D. Baenish, and M. R. Aaron. 1988. *Food versus biomass fuel: Socioeconomic and environmental impacts in the United States, Brazil, India, and Kenya.* Advanced Food Research 32: 185-238.
30. Pimentel, D., and C. W. Hall, editors. 1989. *Food and natural resources.* Academic Press, San Diego, California.
31. Pimentel, D., J. Allen, A. Beers, L. Guinand, R. Linder, P. McLaughlin, B. Meer, D. Musonda, D. Perdue, S. Poisson, S. Siebert, K. Stoner, R. Salazar, and A. Hawkins. 1987. *World agriculture and Soil Erosion.* BioScience 37: 277-283.
32. Pimentel, D., and M. Pimentel. 1979. *Food, energy and society.* Edward Arnold Publishers, Ltd., London, England. 165 pp.
33. Pimentel, D., T. Culliney, I. Buttler, D. Reinemann, and K. Beckman. 1989. *Ecological resource management for a productive, sustainable agriculture.* In D. Pimentel and C. W. Hall [editors] *Food and Natural Resources.* Academic Press, San Diego, California. pp. 301-323.
34. Pimentel, D., W. Dazhong, S. Eigenbrode, H. Lang, D. Emerson, and M. Karasik. 1986. *Deforestation: Interdependency of fuelwood and agriculture.* Oikos 46: 404-412.
35. Population Reference Bureau. 1988. *World population data sheet.* Washington, D.C.
36. Rachie, K. O. 1983. *Intercropping tree legumes with an-*

nual crops. In P. A. Huxley [editor] *Plant Research and Agroforestry.* International Council for Research in Agroforestry, Nairobi, Kenya. pp. 103-116.

37. Reifsnyder, W. E., and H. W. Lull. 1965. *Radiant energy in relation to forests.* Technicul Bulletin No. 1344. U.S. Department of Agriculture, Washington, D.C. 111 pp.
38. Scott, T. W., B. F. Burt, and D. J. Otis. 1984. *Crop, intercrop, and cover crop systems.* Agronomy Mimeograph No. 84-5. Cornell University, Ithaca, New York.
39. Slesser, M., and C. Lewis. 1979. *Biological energy resources.* Halsted Press, New York, New York.
40. Torres, C. 1983. *Potential contribution of Leucaena hedgerows intercropped with maize to the production of organic nitrogen and fuelwood in the lowland tropics.* Agroforestry Systems 1: 323-333.
41. United Nations Environment Program. 1980. *Annual review.* Nairobi, Kenya.
42. Wen, D. 1989. *Soil erosion and conservation in China.* In D. C. Pimentel [editor] *World Soil Erosion and Conservation.* Cambridge University Press, Cambridge, United Kingdom.

11

Sustaining the Resource Base of an Expanding World Agriculture

B. A. Stewart, R. Lal, and S. A. El-Swaify

In 1800, the world's population had reached about 1 billion. It reached 2.5 billion in 1950, 4.7 billion in 1983 (a 1.9 percent average annual increase), and is projected to be 6.1 billion in 2000 (a 1.6 percent average annual increase) (*35*). Agricultural production, however, increased even faster, doubling between 1950 and 1980. This increased production resulted in about a 5 percent increase in per capita daily calorie consumption. People in developed countries consume about 3,315 calories per capita compared to only 2,180 calories consumed by people in developing countries. About 2,300 calories is generally considered the minimum requirement (*13*).

From the beginning of agriculture until about 1950, increased food production resulted almost entirely from an expanded cropland base. Since 1950, however, the yield per unit of land area for major crops has increased dramatically. Much of the increase in yields is due to increased inputs of energy. Between 1950 and 1985, the farm tractor fleet quadrupled, world irrigated area tripled, and fertilizer use increased nine-fold. Between 1950 and 1985, total energy used in world agriculture increased 6.9 times (a 5.7 percent average annual increase) (*2*).

Irrigation played a particularly important role in the rapid increase in food production between 1950 and 1985. The world's irrigated land in 1950 totaled 94 million hectares (232 million acres), but increased to 140 million hectares (346 million acres) by 1960 (a 4.5 percent average annual increase), to 198 million hectares (489 million acres) by 1970 (a 3.5 percent average annual increase), and to 271 million hectares (669 million acres) in 1985 (a 2.1 percent average annual increase) (*28*). However, the current rate of expansion has slowed to less than 1 percent per year (*4*). Irrigated land accounts for 18 percent of the cultivated land, but produces 33 percent of the world's food. A major reason that Asia can support half of the world's population is that it has two-thirds of the world's irrigated area.

The world population continues to increase, and agricultural production by the year 2000 will have to be 50 to 60 percent (a 2.0 to 2.4 percent average annual increase) greater than in 1980 to meet demand. Demand for food and agricultural production in developing countries will double (*13*). The population growth rate in developing countries peaked in 1965 at 2.4 percent annually. It is currently about 2.1 percent and projected to remain at about the 2 percent level through the year 2000 (*13*). The current population and population growth rates for various geographic regions of the world are shown in table 1 (*1*). At the present rate of growth, the population in those rapid-growth regions will double in less than 30 years.

The continued demand for food and fiber, coupled with the sharp decline in irrigation development, means that nonirrigated cultivated land will account for much of the additional agricultural production in future years. Arid and semiarid regions comprise almost 40 percent of the world's land area and are inhabited by some 700 million people, roughly 15 percent of the total population. Low rainfall areas constitute from 75 to 100 percent of the land area in more than 20 countries in the Near East, Africa, and Asia. In some countries of the Sub-Saharan Region of Africa, food production per capita has declined, alarmingly so, during the past decade (*26*).

The world agricultural community faces a continued challenge in meeting the demands for food and fiber. This is particularly true if the resource base is to be sustained. Agricultural production can be increased in many areas in the short term, but it is far less certain that the productivity in many areas can be sustained.

The Issue of Sustainability

Until recently, sustainability was seldom, if ever, mentioned in agricultural literature. Now, it is one of the most widely used words. High costs of irrigation development, escalating energy costs during the 1970s, public concern over potential negative impacts of fertilizers and pesticides on water supplies, soil erosion, soil compaction and salinity problems, as well as other concerns, have caused many people to question whether present agriculture systems can be sustained.

There are many different concepts of sustainability, but none is generally accepted. Sustainability, to many, conveys the idea of a balance between human needs and environmental concerns. The American Society of Agronomy has defined sustainable agriculture as a system "that, over the long term, enhances environmental quality and the resource base on which agriculture depends; provides for basic human food and fiber needs; is economically viable; and enhances the quality of life for farmers and society." It is important to emphasize that this definition implies that a sustainable system results in enhanced productivity. Ruttan (*31*) proposed, as a guide to research, that the definition of sustainability should include (a) the development of technology and practices that maintain and/or enhance the quality of land and water resources and (b) the improvement in plants and animals and the advances in production practices that will facilitate the substitution of biological technology for chemical technology. The first

Table 1. World population growth by geographic region, 1986. (*1*).

Region	Population (million)	Population Growth Rate (percent)	Annual Increment (million)
Slow Growth Regions			
Western Europe	381	0.2	0.8
North America	267	0.7	1.9
Eastern Europe and Soviet Union	392	0.8	3.1
Australia and New Zealand	19	0.8	0.1
East Asia*	1,263	0.8	18.6
Total	2,322	0.8	18.6
Rapid Growth Regions			
Southeast Asia†	414	2.2	9.1
Latin America	419	2.3	9.6
Indian Subcontinent	1,027	2.4	24.6
Middle East	178	2.8	5.0
Africa	583	2.8	16.3
Total‡	2,621	2.5	65.5

*Principally China and Japan.
†Principally Burma, Indonesia, the Philippines, Thailand, and Vietnam.
‡Numbers may not add up to totals due to rounding.

part of the Ruttan definition is similar to that of the American Society of Agronomy, but the key to the second part of the Ruttan definition is the emphasis on more reliance on biological versus chemical technology.

Sustainability should be considered dynamic because, ultimately, it will reflect the changing needs of an increasing global population. Swindale (*34*) stated that much of the public concern in industrialized countries about sustainability has been generated because of high-input systems. Without them, many claim, it would not be possible to meet the food demands of an increasing world population unless more, and less suitable, land is brought into cultivation. Without a more intensive agriculture, sustainability is certainly not assured. However, there are many disturbing features about highly intensive agriculture. Swindale further states that the use of high-input agriculture in the developing world is increasing. Essentially, however, the international agricultural research centers, supported by the Consultative Group on International Agricultural Research (CGIAR), work toward improving agriculture at a lower level of productivity, using far fewer commercial inputs. Research that emphasizes nutrient recycling; crop residue management; intercropping systems, including agroforestry; and the combination of cereals and legumes contributes to the sustainability of low-input agriculture.

Rodale (*29*) stated that there were seven reasons that sustainability has risen to importance on the farm policy agenda. The first is that there currently is enough food in the world. Surpluses and their timely distribution at an affordable price are more of a problem than shortages per se. While many people are going hungry, it is not because of the inability to produce. Policymakers are wondering if this level of production can be sustained.

The second reason is that nonrenewable resources are vitally important to the operation of the conventional system of agriculture in the United States. Without adequate and low-cost supplies of oil, gas, phosphate rock, and a few other similar resources, U.S. agriculture would not be able to turn out large amounts of low-cost food and fiber. At the very least, there are likely to be erratic shifts in the price and supply of these inputs, and this makes long-term planning difficult and problematic.

Third, the high levels of production are taking a toll on the farm environment. Soil erosion, degradation, and deforestation are widespread. This is particularly true in the developing countries where millions of people are trading soil for food.

The fourth reason, Rodale suggests, is that pollution problems can be traced to agriculture. Increasingly, residues of fertilizers and pesticides flow into surface and underground water supplies. There is a high probability that current agricultural methods will have to be changed to avoid unacceptable environmental damage, and people are concerned about how this will affect the sustainability of agricultural production.

Fifth, world population continues to grow, in some places very rapidly. While there is enough food in total today, there will be a much larger population to feed in future years.

Sixth, society will probably need agricultural production for energy and chemical feedstocks as well as for food. As fossil fuel reserves become more expensive, agriculture may have to bear a heavy, additional burden.

Lastly, Rodale lists the concern for the family farm. He sees personalized agriculture as a central theme in the culture and spirit of countries. Even if large industrialized farms are able to produce enough food and fiber, Rodale ponders whether the idea of the good life in rural areas can be sustained without a vibrant family farm sector.

Rodale's seven reasons for the high interest in sustainability involve technical, economic, and social conditions. These are all highly important and interrelated because it may be entirely possible to have technically sustainable agriculture that is neither economically sustainable nor socially acceptable. Herein, however, we will deal primarily with sustainability in regard to how agricultural practices and systems affect the sustainability of the natural resource base.

Hornick and Parr (*19*) reported that, for most agricultural soils, degradative processes and conservation practices occur simultaneously.

The relationship of soil productivity to soil degradative processes and soil conservation practices is illustrated in figure 1. As soil degradative processes proceed and intensify, there is a concomitant decrease in soil productivity. Conversely, soil conservation and reclamation practices tend to increase soil productivity. Therefore, the productivity level of an agricultural soil at any time is a result of the interaction of degradative processes and conservation/reclamation practices shown in figure 1. In natural ecosystems, productivity and sustainability are achieved through the efficient but delicate balance between all necessary in-

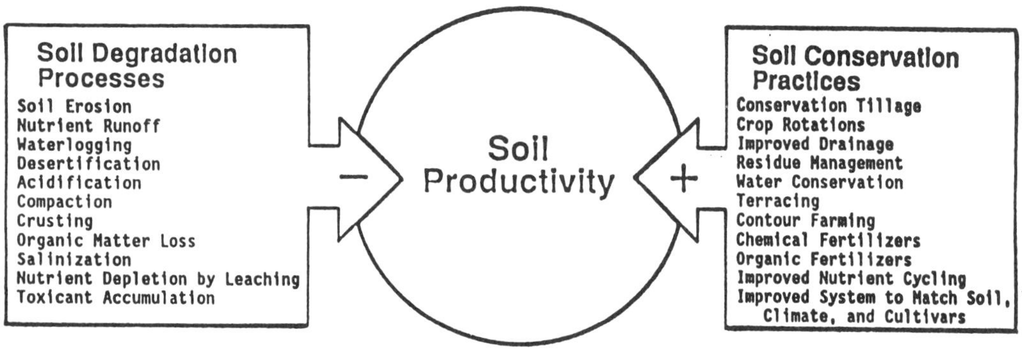

Figure 1. Relationship of soil productivity to soil degradative processes and soil conservation practices Adapted from Hornick and Parr (*19*). Last two processes and practices were added.

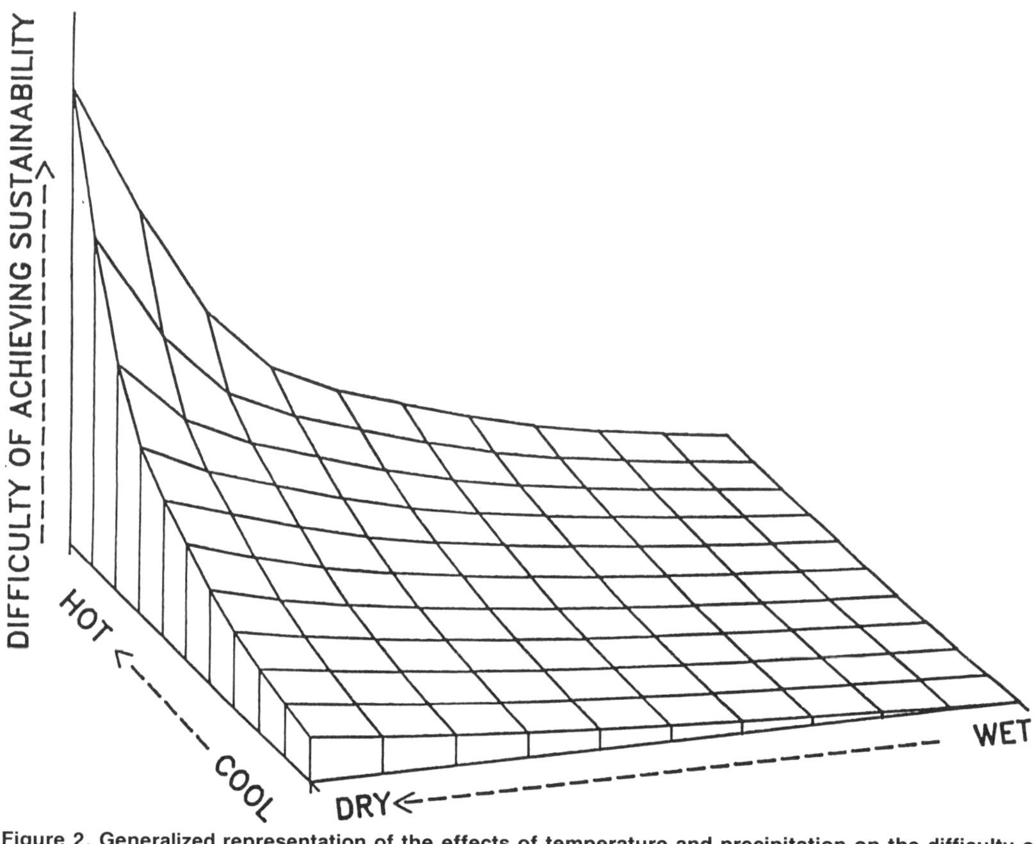

Figure 2. Generalized representation of the effects of temperature and precipitation on the difficulty of developing sustainable agricultural systems.

puts and outputs. Recent efforts, for example, alley cropping in the tropics, which will be discussed later, attempt to duplicate in cropping systems the achievement of the balance that occurs in natural ecosystems. Failure of conventional cropping systems to maintain this balance will ultimately lead to yield gaps.

Using the Hornick and Parr (*19*) concept, a sustainable system is any system in which the benefits from soil conservation practices are equal to or greater than the negative effects of the soil degradative processes. Practices that are generally considered positive can, in some instances, result in soil degradation. For example, the use of organic wastes increases organic matter, improves soil structure, enhances soil water storage, and reduces erosion. Under some circumstances, however, use of organic wastes could result in toxic accumulations or nutrient depletion caused by increased leaching. The important point illustrated by this concept is that degradative processes and soil improvement processes always occur simultaneously, and the net result can be positive or negative.

The negative impacts of soil degradative processes can become very dominant. In extreme cases, soil productivity can be reduced to zero. On the other hand, soil conservation practices rarely improve the productivity of cropland beyond the level experienced initially after virgin land is cultivated. However, soil restorative practices, to be economically feasible and forward-looking, must, of necessity, be focused on improving the soil productivity beyond the level experienced initially after virgin land is cultivated. Examples of such practices include adding fertilizers in areas where essential plant nutrients are deficient and installing water conservation practices to significantly reduce runoff. The most notable increases in dryland productivity generally occur when irrigation practices are instituted or when salt-affected areas are reclaimed.

Development of Sustainable Agriculture Systems

The concept of Hornick and Parr (*19*) presented in figure 1 illustrates that the soil productivity of an agricultural system is dynamic. It changes as a result of the relationships between the negative and positive processes that occur simultaneously. A truly sustainable cropping system is one in which the effects of conservation practices equal or exceed the effects of degradative processes. This concept is equally valid for low-input and high-input systems.

Climatic Effect. Climate is often the most critical factor determining the sustainability of agricultural systems. A generalized view of the effect of varying temperature and moisture regimes on the difficulty of achieving sustainability in an agricultural system is presented in figure 2. As temperatures increase and the amounts of precipitation decrease, the development of sustainable cropping systems becomes more difficult. The reasons for these effects are readily apparent when the processes and practices presented in figure 1 are analyzed.

Perhaps the most dominant soil degradative processes are soil erosion and organic matter decline. As temperatures increase, organic matter decline, particularly in frequently tilled soils, is greatly accelerated. The potential for wind erosion especially and water erosion also generally increases in warmer areas. These same degradative processes are accelerated as the moisture regimes become more arid because, initially, there is a lower inherent organic matter level and less natural vegetation to prevent erosion.

Not only do the soil degradative processes accelerate under hot and arid climatic regimes, the benefits that can be derived from soil conservation practices in these regimes are fewer than those in the cooler and wetter areas. For example, the most important soil conservation practices to alleviate organic matter loss and control soil erosion usually involve crop residues, but the availability of residues decreases sharply in hot and arid areas. Consequently, the negative processes shown in figure 1 can easily become much greater than the positive benefits, and soil productivity can drop precipitously.

The relationships presented in figure 2 do not apply to all climatic regimes. This paper focuses primarily on the drier regions, for which figure 2 was designed. Soil degradation processes also accelerate under very high precipitation regimes, due mainly to erosion, nutrient depletion by leaching, and acidification. Cold conditions severely limit the choice of cropping systems and can also result in waterlogging. Therefore, the shape of the lines in figure 2 would change drastically if extended to very cold or exceedingly high rainfall areas.

Soil Effect. The soil is also an important factor relating to the development of sustainable agricultural systems. The generalized representation shown in figure 2 holds for all soils, but the potential rate and extent of soil productivity decline depend upon soil type. Pierce and associates (*27*) developed a productivity index (PI) for assessing the effect of erosion on soil productivity. Briefly, the PI is predicated on the assumption that soil is a major determinant of crop yield because of the environment it provides for root growth. They evaluated the relative productive potential of soil by calculating a PI based upon the soil's available water capacity, resistance to root growth and development (bulk density), and adequacy of

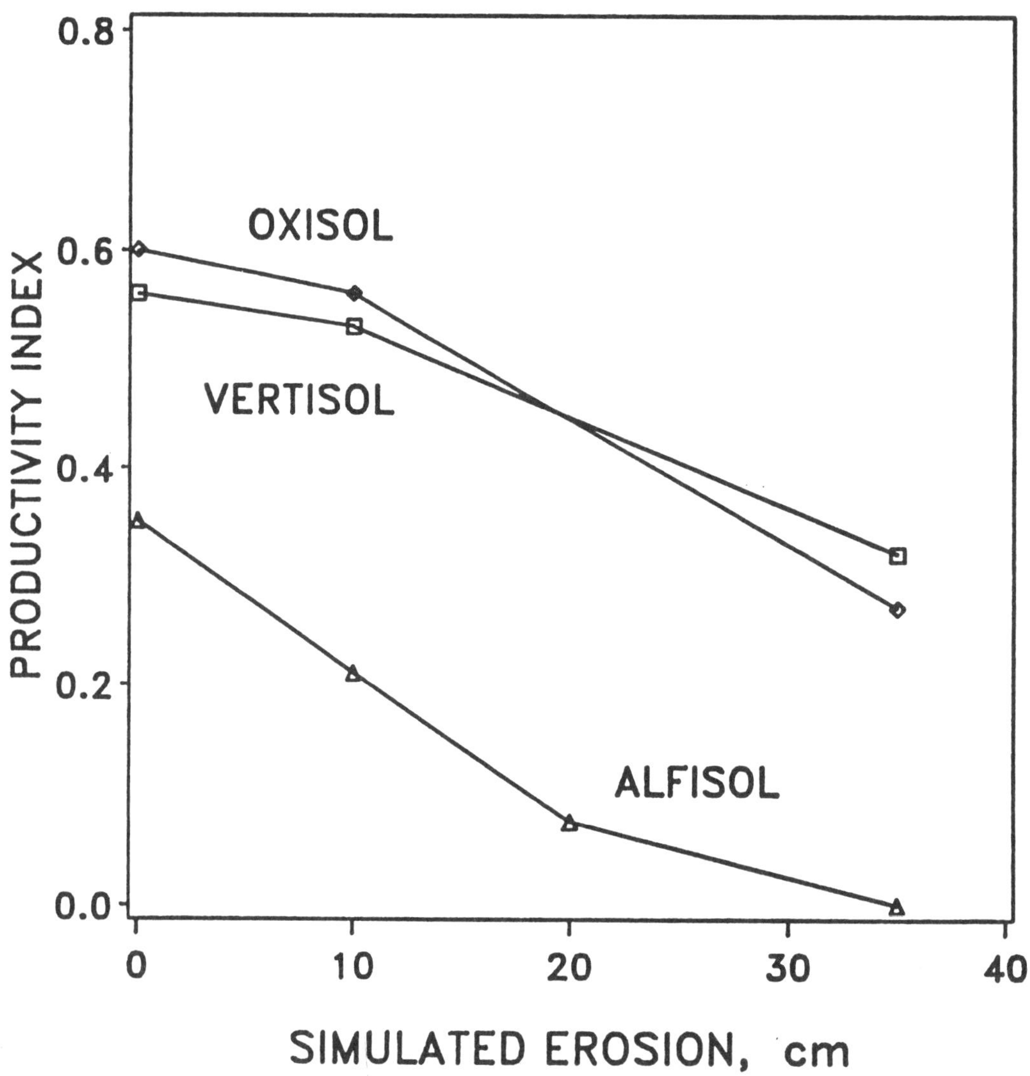

Figure 3. Soil erosion's effects on the productivity of selected soils, as estimated by the Productivity Index model (7).

Table 2. Soil orders and their principal diagnostic properties.

Alfisols	Mineral soils; relatively low in organic matter; relatively high base saturation; an illuvial horizon of silicate clays; moisture adequate to mature a crop in most years.
Aridisols	Mineral soils; relatively low in organic matter; inadequate moisture to mature a crop without irrigation in most years; some pedogenic horizons.
Entisols	Mineral soils; weak or no pedogenic horizons; no deep, wide cracks in most years.
Histosols	Organic in more than half of upper 80 cm.
Inceptisols	Mineral soils; some pedogenic horizons and some weatherable minerals; moisture available to mature a crop in most years; no horizon of illuvial clays; relatively low in either organic matter or base saturation, or in both.
Mollisols	Mineral soils; thick dark surface horizon, relatively rich in organic matter; high base saturation throughout; no deep, wide cracks in most years.
Oxisols	Mineral soils; no weatherable minerals; inactive clays; no illuvial horizon of silicate clays.
Spodosols	Mineral soils; an illuvial horizon of amorphous Al and organic matter, with or without amorphous Fe.
Ultisols	Mineral soils; an illuvial horizon of silicate clays; low base saturation; moisture adequate to mature a crop in most years.
Vertisols	Clayey soils; deep, wide cracks at some time in most years.

pH to a depth of 100 cm (39 inches). El-Swaify (*7*) applied the PI concept to the major soil orders of the semiarid tropics (Figure 3). He found that Alfisols are much more vulnerable than Oxisols or Vertisols to the impacts of erosion. The topography, specifically the slope component, can also significantly affect soil degradation, particularly runoff and erosion, but cropping systems or engineering practices can often be employed to control these processes.

All soils can be classified in 10 orders, according to *Soil Taxonomy* (*32*). The orders are differentiated by the presence or absence of diagnostic horizons or features that mark differences in the degree and kind of the dominant sets of soil-forming processes that have occurred. A brief description of the orders is given in table 2. The distribution of the soil orders in the world and selected regions is shown in table 3.

Lal (*23*) discussed soil resources of the tropics. As shown in table 3, major soils of the tropics are Oxisols, Aridisols, Alfisols, Ultisols, Entisols, and Vertisols. Much like soils in the temperate zone, soils in the tropics are highly diverse and variable, depending upon geology, geomorphology, vegetation, and rainfall patterns. These soils can be broadly grouped into three major categories: (1) soils containing predominantly low-activity clays, (2) soils of mixed mineralogy and those containing high-activity clays, and (3) soils of volcanic origin. Soils of the rainforest and semideciduous forest regions are those containing low-activity clays. These soils are characterized by clay fractions mostly composed of kaolinite and halloysite, with hydrous oxides of iron and aluminum. Alfisols, Ultisols, and Oxisols are of this category and comprise about 50 percent of the land surface area of the tropics and about 27 percent of the world land area.

Soil orders containing high-activity clays occur mostly in semiarid and arid regions. They include Aridisols, Vertisols, Mollisols, Inceptisols, and Entisols. The predominant soils of the arid tropics, Aridisols and Vertisols, occupy about 20 percent of the tropical land area. Recently developed Inceptisols and Entisols are relatively fertile and occupy about 16.5 percent of the tropical land area. In general, soils of high-activity clays are less easily dispersed than those of low-activity clays.

Lal (*23*) characterized the general soil constraints to intensive use of uplands for some of the major soil orders as shown in table 4. This information, used in conjunction with the generalized climate effect presented in figures 2 and 3, allows a preliminary assessment of the sustainability at a given location.

El-Swaify and associates (*10*) summarized extensive research studies conducted on contrasting soils under the same climatic conditions. Vertisols and Alfisols are both extensively represented within the experimental farm at the International Crops Research Institute for the Semi-Arid Tropics (ICRISAT Center) located near Hyderabad, India.

Vertisols are deep, black soils, generally called black cotton soils. They are abundant in India, Sudan, Ethiopia, Australia, the United States, and several other countries. Vertisols are heavy in texture; by definition, they contain more than 30 percent clay-sized particles. Texture is more or less uniform throughout the profile. Still, deeper layers may be higher in clay content than surface layers. Vertisols may be derived from a variety of parent materials, such as basalt, shales, limestone, volcanic rocks, tuffs, gneisses, schists, and basic alluvium or colluvium materials (*24*). Most of the soil clay is of the swelling, 2:1 lattice type; montmorillonite is often reported as the dominant clay mineral.

Vertisols are physically quite responsive to changes in the chemical environment; the effects of exchangeable sodium depend upon the level of total salts in the soil (*11*). Cultivation practices on Vertisols are particularly affected by the sticky nature, poor infiltration, and impeded internal drainage of the soils while wet, as well as excessive hardness and difficult workability while dry. Thus, cultivation prac-

Table 3. Land area of the soil orders in the world, United States, Tropics, and the Semiarid Tropics of Africa, Latin America, and Asia.*

				Semiarid Tropics		
	World	*United States*	*Tropics*	*Africa*	*Latin America*	*Asia*
	million ha					
Alfisols	1,730	125.0	800	466	107	121
Aridisols	2,480	107.0	900	440	33	47
Entisols	1,090	74.0	400	255	17	—
Histosols	120	4.8	—	—	—	—
Inceptisols	1,170	170.0	400	38	—	28
Mollisols	1,130	232.0	50	—	78	—
Oxisols	1,120	0.1	1,100	188	—	—
Spodosols	560	47.0	—	—	—	—
Ultisols	730	120.0	550	24	8	20
Vertisols	230	9.1	100	51	—	80
Mountains, streams, lakes, etc...	2,810	54.0	600	—	70	23

*From Soil Survey Staff (*32*); Lal (*23*), and El-Swaify and associates (*10*). Because various sources are used, the sums may not agree with world totals.

Table 4. General soil-related constraints to intensive use of upland soils in the tropics 23).

	*Soils Constraints**					
Soil-related Constraint	*Oxisols*	*Ultisols*	*Alfisols*	*Inceptisols*	*Vertisols*	*Mollisols*
Erosion	2	2	3	3	3	1
Compaction	3	3	3	2	2	2
Crusting	1	1	3	3	2	1
Drought	2	2	3	1	2	1
Shallow effective rooting depth	3	3	3	1	2	1
Trafficability	1	1	1	1	3	2/3
Supraoptimal soil temperatures	2	2	3	3	2	2
Chemical/nutritional	3	3	2	1	2	2/3†

*3 represents severe; 2 represents moderate; and 1 represents slight.
†Alfisols and Mollisols in temperate zones are generally more fertile.

Table 5. Grain yields from a maize/pigeonpea intercrop system and a maize/chickpea sequential system compared with traditional rainy season fallow on a deep Vertisol in an operation-scale watershed at ICRISAT Center (*10*).

	Grain Yield (Mg/ha)				
Cropping System	*1976-1977*	*1977-1978*	*1978-1979*	*1980-1981*	*Mean*
Maize/pigeonpea intercrop system					
Maize	3.29	2.81	2.14	2.92	2.79
Pigeonpea	0.78	1.32	1.17	0.97	1.06
Maize-chickpea sequential system					
Maize	3.12	3.34	2.15	4.18	3.20
Chickpea	0.65	1.13	1.34	0.79	0.98
Traditional fallow and single post-rainy season crop					
Chickpea	0.54	0.86	0.53	0.60	0.63
Sorghum†	0.44	0.38	0.55	0.56	0.48

*Available water in the deep vertisols is 230 mm.
†(*Sorghum bicolor* L.).

tices must coincide with the specific range of soil water contents at which the soil is trafficable and of a consistency that allows easy preparation of the land and the production of good tilth. Vertisols are highly susceptible to erosion, even though the values for the universal soil loss equation's soil erodibility factor, K, for many Vertisols have averaged 0.3 on a scale of 0 to 0.69 for the lowest and highest observed erodibility, respectively (*9*).

Alfisols, the most abundant soils in the semiarid tropics, cover nearly 16 percent of the tropics and 33 percent of the semiarid tropics. They occur extensively in southern Asia, western and central Africa, and many parts of South America, particularly northeastern Brazil (*10*). Alfisols are derived mostly from granites, gneisses, and schists, but occasionally, they are derived from sandstone, mica, acid trap, quartzite, and shale.

Characteristically, Alfisols possess an argillic horizon within the profile. Thus, the clay content of these soils increases with depth, although shallow and gravelly Alfisols are also common as a result of erosion. The enrichment of surface layers with coarse particles is assumed to be the result of clay migration with percolating water, termite activity, and/or selective removal of fine particles by soil erosion. Effective rooting depths of crops are restricted either by the limited soil depth or the compact argillic horizon, which may restrict water and root penetration. Restricted root development on these soils prevents many crops from withstanding even moderate droughts; poor crop yields are frequent.

Alfisols are characterized by lack of structural development. This is due to the combined contributions of low contents of clay-sized particles, particularly at the surface, inactivity of the prevailing clay minerals (kaolin, with varying, but small proportions of 2:1 clays and sesquioxides), and, with the prevailing cropping systems in the semiarid tropics, the tendency to stabilize only low amounts of decomposed organic matter within the soil mass. A major consequence of the lack of aggregation is the tendency of these soils to display rapid surface sealing following rainfall, crusting with subsequent drying cycles, and setting hard on ultradesiccation.

From the chemistry and fertility standpoint, Alfisols are generally quite low in fertility, but this depends on their clay content, mineralogy, and organic matter content. Because of the generally low cation exchange capacities, organic matter plays an important role in determining the nutrient status of Alfisols. The lack of active clays also indicates that organic matter must play an important role in controlling soil structural characteristics.

The erodibility of Alfisols varies widely; reported values for the universal soil-loss equation's erodibility value range from less than 0.1 for soils in West Africa (*30*) to 0.4 in other regions, indicating that the soils are quite susceptible to erosion by water (*8*).

El-Swaify and associates (*10*), in summarizing the many studies conducted on the contrasting Alfisols and Vertisols at the ICRISAT

Center, showed the importance of the soil in establishing sustainable agricultural systems. Improved management of the Alfisols for conventional cropping succeeded in increasing crop yields over traditional management systems, but effective practices for improved soil and water conservation have not been established. This is primarily because of the extreme structural instability of these soils. Improved structure will likely not be achieved without increasing soil organic matter, and this is indeed difficult to do in a semiarid climate, as illustrated in figures 2 and 3.

Improved management of the Vertisols was much more successful. The recommended technological improvements had immediate and dramatic benefits over the traditional management systems. Traditionally, farmers cultivate these soils only during the post-monsoon season, a practice that carries little economic risk, but leaves the soil bare and unprotected during the monsoon. Improved watershed-based management approaches allow both rainy season and post-rainy season cropping of Vertisols in assured rainfall areas. Crop yields were increased several-fold (Table 5). In addition, runoff and erosion were reduced significantly. On experimental watersheds with a 0.4 percent slope, during the period from 1975 to 1980 (with a mean annual rainfall of 818 millimeters, 32.2 inches), total runoff averaged 220 millimeters (8.66 inches), and soil loss averaged 6.64 megagrams per hectare (14.87 tons per acre) for traditional farming, compared to 91 millimeters (3.58 inches) and 0.86 megagrams per hectare (1.93 tons per acre) for the improved management system.

There are many other examples that can be cited to illustrate the concepts presented in figures 1 and 2. One is the infamous Dust Bowl that occurred in the 1930s in the Great Plains of the United States. The agricultural systems in this region at that time were largely developed by farmers moving from the Corn Belt region of the United States to the Great Plains, and taking with them their tillage implements and experience. The Corn Belt is cool and humid. In contrast, the Great Plains is significantly drier and the temperature regime ranges from cool to warm. The farming practices that had been successful in the Corn Belt for developing sustainable cropping systems led to crop failures and the most severe erosion problems ever witnessed on the North American continent. The degradative processes, namely soil erosion and organic matter loss, were simply so much greater than the conservation practices that soil productivity dropped to levels that could no longer support commercial farming operations. Many farms were abandoned and the future of the Dust Bowl area was in question. Many forecasters thought the area had no agricultural future. However, much has been learned during the more than 50 years since that ecological disaster. The focus has been on developing soil conservation practices that affect soil productivity positively enough to largely offset the ever-present soil degradative processes.

The first practice that had to be replaced was clean tillage. Clean tillage generally increases organic matter decline and also greatly increases erosion potential. Water conservation practices and residue management systems that emphasize the maintenance of residues on the soil surface have led to cropping systems in the Great Plains that are clearly sustainable. The challenge has been much greater in the southern Great Plains than in the northern Great Plains because of the higher temperature regime in the southern area. A review of the processes and practices in figure 1 clearly indicates that the negative effects increase with temperature, while the opportunities and benefits from conservation practices tend to decrease.

The fact that there are areas in the United States where agricultural systems are not sustainable was clearly recognized and addressed by the 1985 Food Security Act passed and implemented by the United States Congress. Provisions contained in Title XII of the act provided some of the most far-reaching changes in soil and water conservation programs in the United States since the 1930s. For the first time, to continue to receive U.S. Department of Agriculture (USDA) program benefits, farmers are required to follow conservation farming practices on highly erodible land. Two programs relate directly to highly erodible land. One is the Conservation Reserve Program that encourages farmers to stop growing annually tilled crops on highly erodible cropland for a 10-year contract period. The goal of USDA is to convert, by 1990, up to 18 million hectares (45 million acres) of highly erodible cropland to perennial vegetation. Farmers submit bids to enter erodible cropland into the program, and if accepted, they receive an annual rental payment for each year of the contract period, together with approximately 50 percent cost-sharing for the establishment of the required cover of grass or trees. The definition of highly erodible cropland for purposes of the Conservation Reserve Program is based on theoretical estimations of actual erosion rates. To be eligible for the program, land must be calculated to have undergone erosion from 1981 to 1985 at an annual rate greater than three times the estimated rate at which productivity can be maintained [15 to 37.5 metric tons per hectare (7 to 17 tons per acre) depending on the specific soil], or eroding at a rate of two times the estimated rate and with significant gully erosion.

The second program deals with the conservation compliance provision of the act. This provision requires farmers who continue to crop highly erodible land to have a conservation plan approved and to begin actively implementing that plan by January 1, 1990, to be eligible to receive USDA program benefits. The plan must be fully implemented by January 1, 1995. Highly erodible cropland for this law have been defined as those with potential erosion of more than eight times the estimated rate at which the soil can maintain continued productivity. The potential erosion is based on a field bare of vegetation and without any conservation practices. Fields are considered highly erodible where one-third or more of the soils are highly erodible or where the field contains 20 hectares (51 acres) or more of highly erodible land. More than 25 percent of the cropland in the United States is subject to the law. Conservation plans, for purposes of this law, require that actions must be taken to bring about ''significant reductions'' in erosion rates.

Although it is too early to make judgments about the success of this program, this is a clear example of how government policy can have a major impact on managing the natural resource base. The concepts behind the Conservation Reserve Program and the conservation compliance provision are quite sound. However, there is a lot of ambiguity in applying quantitative measurements. This illustrates the need for a systematic productivity- and sustainability-based means of land use planning.

To date, farmers have placed about 13.8 million hectares (34 million acres) of cropland into the Conservation Reserve Program. More than 60 percent of this land is located in the Great Plains. There is concern among many conservationists that at the end of the 10-year period farmers will return Conservation Reserve Program land to cropland and the soils will again be degraded. Programs may be instituted that will encourage landowners to allow these acres to remain in perennial grasses or to use the land for livestock production. However, if economic conditions are favorable, at least a portion of this land will be returned to crop production. Research is needed to determine whether it is feasible to return this land to crop production using only limited- or no-till systems This could greatly slow the rate of decline in soil organic matter, making the system more sustainable over a longer period.

Without government policies that consider both the interests of the farmer and society as a whole, land degradation problems will intensify greatly as population pressures increase. In many areas, it may very well be necessary to periodically remove land from crop production for several years to allow regeneration of the resource base. The Conservation Reserve Program discussed above is perhaps one way this could be accomplished. The land presently in the program could be returned to annual cropping until certain thresholds of degradation are reached and then be returned to the program for regeneration. This ''in-and-out'' approach would be similar to the traditional farming system that was practiced for centuries in Africa. That practice is generally referred to as bush fallow. It consists of clearing the brush from the land and then growing crops for 2 to 3 years. Then, the land is allowed to go back to its native vegetation for 7 to 9 years or longer, and additional land is cleared for crop production.

This system was clearly sustainable and was simply a matter of allowing the processes and practices listed in figure 1 to alternate as the dominant factors. During the cropping years, the degradative processes were dominant, the organic matter decline is rapid, and soil structure deteriorates significantly. Hartmans (*17*) asked, "Why do most tropical soils become unproductive and useless after only a few years, and what can be done to arrest this deterioration?" He stated that results of 15 years of research at the International Institute of Tropical Agriculture (IITA) at Ibadan, Nigeria, in pursuit of these questions are quite clear. Chemically, the land becomes more acid very rapidly, and essential plant nutrients are lost. Physically, the soil seems to collapse on itself. It becomes more dense, and erosive forces often cause the finer particles to disappear, leaving a sandy or gravelly material. The soil loses its capacity to form stable aggregates because the binding material, the soil organic matter, is gone. The result is a rapid downward spiral of soil productivity.

Again, referring to figure 1, the degradative processes are simply dominating the system, resulting in a rapid loss of soil productivity. The bush fallow system, which cycled land in and out of cropland, allowed the degradative processes and soil conservation practices to cycle as the dominant factors affecting soil productivity. This is in direct contrast to cropland management in temperate areas, like the U.S. Corn Belt and Western Europe. In those areas, degradative processes and soil conservation practices tend to offset each other, and, under good management, soil productivity cannot only be sustained but also enhanced.

The IITA has developed a number of approaches to prevent or delay the decline of soil productivity in the tropics. The approach is to imitate, as closely as possible, the traditional bush fallow system. No-till systems have been developed and found effective under some conditions. However, Hartmans (*17*) stated that, in the tropics, no-till was often difficult in practice and hard to adopt on a large scale. Weeds and bush, unchecked by plowing, compete with crops. Lal (*22*) did develop successful no-till systems, but a high level of management is required. Ecological limits of application of no-till farming can be increased by developing appropriate cropping/farming systems based on cover crops, cropping sequence, agroforestry techniques, etc.

Alley cropping is a system that shows great promise in many areas for imitating the natural system. Alley cropping uses rows of fast-growing trees between broad beds of field crops to develop conditions similar to the recycling system of the original forest conditions. Leucaena and Glyricidia, leguminous tree species, have been used successfully. The shrubs are pruned at regular intervals during the growing season and the clippings used for organic mulch. The shrubs also provide some shade for temperature control and are sufficiently deep-rooted to continue growth during the dry seasons in some areas.

Alley cropping is a labor-intensive system. It can be adapted in ecological regions with rainfall regimes of 1,200 to 2,000 millimeters per year (47 to 78 inches per year). For areas with less than 1,000 millimeters (39 inches) of rainfall, perennial shrubs compete for water and nutrients. For acidic soils in high rainfall zones, Leucaena and other shrubs are not adapted; all are prone to insects and pests.

Unless alley cropping is proven economical in the context of specific socioeconomic constraints, it is not going to be adapted by the farming community. Evaluating socioeconomic aspects of alley cropping involves assessment of all components, especially labor. Additional labor is required for managing perennial hedgerows, including seeding, pruning, incorporation of prunings in the soil, etc. To date, there is little economic validation of these agroforestry systems. It is really a question of growing nitrogen with additional labor rather than purchasing nitrogen. In additon, social acceptability can also be an obstacle due to labor demands and the introductioon of exotic species. Considering all factors, returns per unit of labor is the driving force in widespread

Table 6. Potential benefits of alternative land use systems compared with conventional, annual, sole-cropped systems in Alfisols of the semiarid tropics (7).

	Potential Benefits*					
Alternative Land-Use System	*Improved Water-Use Efficiency and Access to Available Resources*	*Maintained and Enhanced Fertility*	*Reduced Surface Sealing and Crusting, and Lowered Soil Temperature at Sowing*	*Decreased Runoff and Soil Erosion*	*Increased Soil Structural Stability*	*Reduced Pest and Disease Incidence*
Conventional annual cropping systems						
Sequential cropping	-	-	?	?	?	?
Relay cropping	-	-	?	+	?	?
Ratoon cropping	-	-	?	?	?	?
Intercropping	+	?	?	+	?	+
Non conventional annual cropping systems						
Minimum Tillage	+	+	+	+	+	-
Intensive deep tillage	+	?	+	+	?	+
Agroforestry systems						
Grass and legume ley farming	+	+	+	+	+	+

* + indicates that the alternative land use system performs better than conventional, annual sole cropping;
- indicates that the alternative land use system is technically inferior to annual sole cropping;
?indicates that the impact of the alternative land use system compared with that of annual sole cropping is not clearly known, but is probably not very different.

adoption of this technology.

From the sustainability perspective, alley cropping and agroforestry in general is needed to provide a stabilizing perennial element of vegetation to the agroecosystem. This is critical for regions with drought-prone soils and soils with marginal capacity for water storage, such as Alfisols. Other cropping systems, for example, pastoral systems and legume-ley farming, also impart sustainability. The potential benefits of alternative land use systems, compared with conventional, annual, sole-cropped systems grown on Alfisols in the semi arid tropics, are illustrated in table 6.

Figure 2 is a generalized representation of the effects of climate on the difficulty of achieving a sustainable cropping system and cannot be used for quantifying specific locations. But some relative comparisons of locations are shown in figure 4. The points for the various locations were obtained by plotting the annual mean temperature, annual mean precipitation, and the temperature/precipitation value. These index values are shown only for a general comparison; they should not be considered in any way as representing a specific quantitative difference. The values do, however, allow some relative comparisons to be made. For example, the precipitation from north to south in the Great Plains of the United States and Canada is fairly uniform, but the temperature increases significantly from north to south. Conversely, moving east from the U.S. northern Great Plains to the Corn Belt, the temperature remains somewhat similar, but precipitation increases substantially. Comparing the index values in figure 4 for Saskatoon, Saskatchewan (*2*); Mandan, North Dakota (*22*); Akron, Colorado (*16*); Bushland, Texas (*11*); and Big Spring, Texas (*9*) suggests that the difficulty of achieving sustainable agricultural systems increases significantly as a result of increasing temperature. Likewise, comparing Ames, Iowa (*24*), and Akron, Colorado (*16*), illustrates the effect of somewhat constant annual mean temperatures, but decreasing precipitation increases the difficulty of achieving a sustainable system at Akron.

The nitrogen and carbon contents of selected soils from the Great Plains of the United States are presented in table 7 for northern, central, and southern locations. The values at the time of initial cultivation of the soils were considerably higher in the northern areas, and they remained significantly higher after approximately 35 years of cropping. The organic carbon and nitrogen values for the southern-area soils decreased to very low levels with cropping, making these soils extremely vulnerable to wind and water erosion and reducing their water-holding capacities. This again illustrates the increasing difficulty of achieving sustainable systems as annual mean temperatures increase at somewhat constant precipitation levels. This effect is compounded when precipitation decreases at the same time that temperature increases. A somewhat extreme example for such a situation occurs in Tombouctou, Mali (Figure 4), where the annual mean temperature is 28° Celcius and the annual precipitation is 196 millimeters (7.7 inches). As suggested by the "index" in figure 4, it will likely be extremely difficult to develop a sustainable agricultural system under such climatic extremes, and particularly so if the soil is an Alfisol or some other soil that has significant constraints (Table 4).

International Soil Management

There is increasing concern and attention about managing natural resources, particularly soil and water. A major reason for this focus has been the decline in agricultural production in many parts of Africa. Another is the apparent global climate change due to the greenhouse effect, caused partly by soil degradation. Agricultural production declined during the 1970s in 13 African states, and it grew by less than 2.5 percent in 15 others. Consequently, per capita food production has been steadily falling in these countries (*14*).

The increasing pressure on natural resources in these countries has resulted in desertification of some areas. Desertification, until recently, was generally considered to be primarily the result of prolonged drought. However, it has become apparent in recent years that human misuse of land is perhaps the basic factor, and drought is only an exacerbating factor (*6*). Dregne states that poverty, ignorance, and greed are the indirect causes of desertification. The direct cause is mismanagement of the land through such prac-

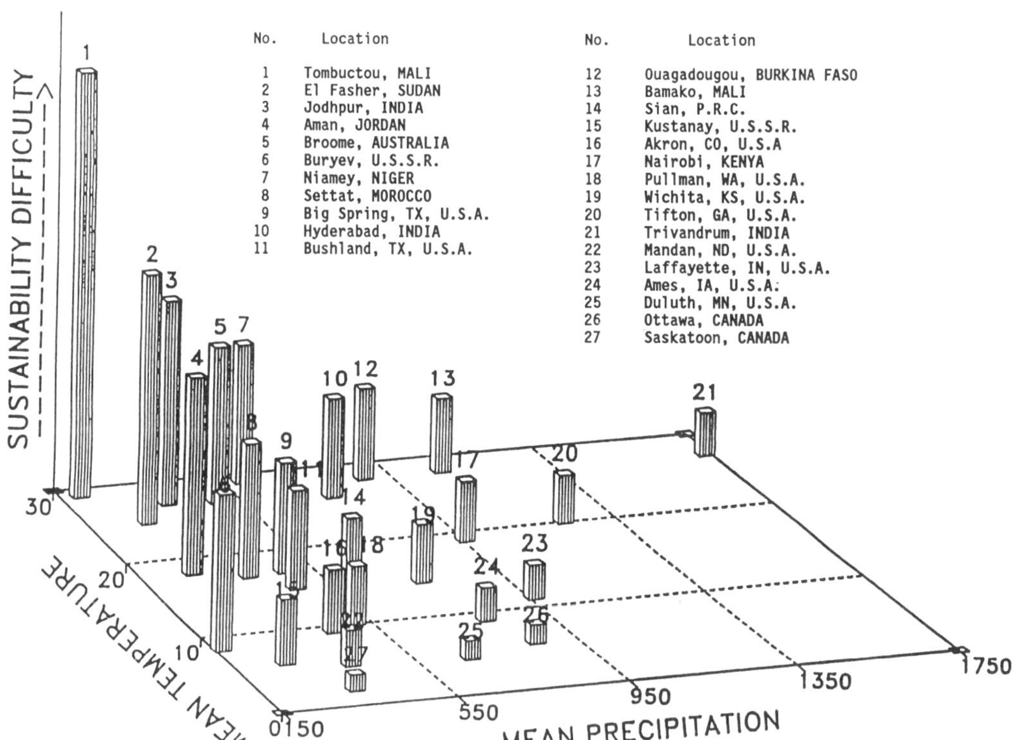

Figure 4. Mean annual temperature and precipitation for various locations and temperature/precipitation as an "index" indicating difficulty of developing a sustainable agriculture system.

Table 7. Nitrogen and carbon content in virgin and cropped soils for selected soils in the U.S. Great Plains (*16*).

	Mean Temperature (C°)	Mean Precipitation (mm)	Texture	Virgin Organic-C (%)	Cropped Organic-C (%)	Virgin Soil-N (%)	Cropped Soil-N (%)	Cropped (years)
Mandan, North Dakota	5.0	401	Fine Sandy Loam	2.11	1.45	.160	.116	30
Dickinson, North Dakota	4.6	397	Loam	3.64	1.51	.293	.149	40
Havre, Montana	5.6	292	Clay Loam	1.75	.83	.150	.090	31
Moccasin, Montana	5.7	384	Clay Loam	3.24	2.19	.300	.205	39
(mean)	5.2	369		2.69	1.52	.226	.140	35
Akron, Colorado	9.3	442	Silt Loam	1.42	.77	.134	.086	39
Colby, Kansas	10.6	469	Silt Loam	1.83	1.01	.165	.105	41
Hays, Kansas	12.2	587	Silty Clay Loam	2.47	1.21	.220	.122	43
Garden City, Kansas	12.1	458	Fine Sandy Loam	1.13	.69	.120	.077	39
(mean)	11.1	489		1.71	.92	.160	.098	41
Woodward, Oklahoma	14.5	604	Fine Sandy Loam	—	—	.080	.032	33
Lawton, Oklahoma	16.3	729	Silt Loam	—	—	.154	.074	31
Dalhart, Texas	12.4	460	Fine Sandy Loam	.72	.44	.067	.042	39
Big Spring, Texas	17.2	479	Fine Sandy Loam	—	—	.060	.041	41
(mean)	15.1	569		0.72	0.44	.090	.047	36

tices as overgrazing, tree cutting, improper tillage methods, poor water distribution systems, and overexploitation. This means that desertification control is a matter of natural resource management, with emphasis on cropland and rangelands. Dregne (*6*) lists the principal desertification processes as (1) degradation of vegetative cover, (2) water erosion, (3) wind erosion, (4) salinization, (5) soil crusting and compaction, (6) reduction in soil organic matter, and (7) excessive toxic substances in the soil. As the list indicates, desertification is simply another word for land degradation, but in an arid land context. Dregne's processes (*6*) are essentially the same as those presented and discussed earlier in figures 1 and 2.

Six well-established trends were identified by the United Nations Food and Agriculture Organization (FAO) as the causes of the agricultural crisis in Africa: (1) a bias against agriculture in government policy; (2) high population growth rates; (3) a decline in the rate at which arable land and harvested area are being developed; (4) a lack of technological change, which is leading to a widespread stagnation or even decline in crop yields; (5) accelerated degradation of the environment; and (6) an external economic environment that is making it more and more difficult for most African countries to balance their budgets (*14*).

Although the spotlight has been on Africa in recent years, the concern about degradation of soil and water resources is worldwide. The Middle East, India, and China also have major land degradation problems and rapidly increasing populations, although the population growth rate in China has decreased significantly in recent years. Land degradation problems, however, are not confined to developing countries, although they are generally worse in these areas because of the increasing pressures brought about by population growth and lack of resources.

The widespread concern about global environmental change is also being linked to severe problems of soil degradation. Soil degradation affects global climate both directly and indirectly. Directly, soil degradation contributes to an increase in atmospheric carbon dioxide through rapid decomposition of soil organic matter. The problem is particularly severe in the tropics and subtropics where the mean annual temperatures are high. It is important to recognize that world soils contain about three times more reserves of organic carbon than world vegetation, 1,500 billion versus 560 billion metric tons, respectively (*20*). Degradation of soil and decomposition of organic matter are, therefore, potentially very dangerous to global warming. Indirectly, soil degradation undermines the most basic of all resources. Declining agricultural productivity necessitates bringing more land under cultivation through deforestation, burning, and cultivation of marginal and steep land. All of these practices contribute to the greenhouse effect. Controlling the rate of land degradation and restoring the biological productivity of degraded ecosystems may be the crucial factor in reversing the trend in the emission of greenhouse gases.

What Is Our Role?

Our first and perhaps most important role is to properly assess the potential of soil resources and their environment. In recent years, many people have called for a "green revolution" for the drylands similar to the very successful green revolution that occurred in Asia. While we join this cause, any such revolution in dryland areas must be recognized as being vastly different from what occurred in Asia under irrigated or reliable rainfall conditions. In the Asian case, new cultivars provided the key to large yield increases. This led many to believe that new cultivars will also be the key for the less-favored areas. However, while plant breeding will be important, the most important issues in semiarid areas will be to efficiently manage precipitation and prevent degradation of the soil resource base. Too often, unrealistic goals are stated, for example, "Among the major requirements are cultivars characterized by drought resistance, an ability to thrive on highly saline, alkaline, and acid soils, and on soils that contain high levels of aluminum or which are deficient in phosphorus or nitrogen" (*14*).

We must realistically assess the resource base and clearly state, based on the best scientific knowledge available, the potential and limitations of agricultural areas. Buringh and associates (*3*), Norse (*25*), De Vries and De Wit (*5*), and Higgins (*18*) have estimated that agricultural potential in Africa is not a major constraint. The quality of land resources is also adequate. Institutional and socioeconomic constraints are the major obstacles.

Although soil scientists cannot address the social, economic, and infrastructure problems in these areas, the governments and institutions cannot address them adequately without valid information from soil scientists about the resource base. Soil scientists must also realistically address the potential gains in food and fiber production likely to occur in various areas. While there is no question that increased production can occur in many of these areas, the increases cannot be as significant as have occurred in more favored environments because of the lack of precipitation or soil fertility constraints. Also, in the less-favored environments, the potential for reducing yields as a consequence of land degradation can sometimes be as great or greater than the potential for increasing yields. A report by FAO (*15*) stated that 80 percent of the rangeland and 60 percent of the cropland in the developing country drylands was estimated to be declining in productivity.

Soil scientists must also take an active role in technology transfer and in monitoring the effects of these technologies on the soil resource base. As already discussed and illustrated in figures 1, 2, and 3, an effective and sustainable technology in one area may fail miserably after only a few years when transferred to an area where the climate and soil are different.

Steiner and associates (*33*) addressed the issue of improving and sustaining productivity in dryland regions of developing countries. They identified the constraints and proposed strategies for alleviating these constraints. They categorized the constraints into four categories: (1) physical constraints (agroclimatic conditions and soil characteristics); (2) technological constraints (soil fertility, crop germplasm, and production practices); (3) institutional and infrastructure constraints (credit, marketing and distribution, research and technology transfer, fertilizers and pesticides, and farm-level knowledge base); and (4) socioeconomic constraints (population growth, land tenure and fragmentation, role of women, pastoral grazing, lack of labor, and macroeconomic policy constraints). Clearly, soil scientists have a major role to play in addressing these constraints to sustainability.

The focus of this discussion has been on the less-favored agricultural areas because these areas present the greatest challenge in maintaining sustainable agricultural systems. However, we should not lose sight that maintaining sustainability is also a major concern in some of the more favorable climatic zones and in irrigated areas. Keen (*21*) recently reported that of the acres currently irrigated about half of them need improvements or modernization because of degradation of soil and water quality, inefficient use of water by plants, siltation of reservoirs, and decay of the distribution channels. There are also areas where the irrigation water is supplied by groundwater, which is being exhausted. An example is the southern Great Plains of the United States where some previously irrigated land has reverted to dryland farming because the groundwater was depleted by some 40 to 50 years of irrigation.

Research and Development Needs

The development of productive, profitable, and sustainable agricultural systems will require improved understanding of the biophysical system. A more complete characterization of the soil resource—physically, chemically, and biologically—is clearly needed. The climate must also be better understood and used. Even though the climate in many regions is a constraint, it is, in reality, a resource that must be analyzed and used to the fullest extent feasible. This can only be done by a better understanding of probabilities. In the past, too much reliance has been given to monthly averages, which have little true meaning, particularly in low precipitation regions. As annual precipitation decreases, the percentage variability increases, and averages can become very misleading.

The international agricultural research centers supported by CGIAR are playing a major role in the development of agricultural systems in the developing countries. CGIAR is an association of countries, international and regional organizations, and private foundations dedicated to supporting a system of agricultural research centers and programs around the world. There are presently 13 CGIAR-supported research centers. The centers that most directly focus a major part of their programs on the soil resource and climate, as these factors affect sutainable agricultural systems, are the International Institute of Tropical Agriculture (IITA), Ibadan, Nigeria, and the Centro Internacional de Agricultura Tropical (CIAT), Cali, Colombia, for the humid tropics; the International Crops Research Institute for the Semi-Arid Tropics (ICRISAT), Patancheru, India, and the ICRISAT Sahelian Center, Niamey, Niger, for the semiarid tropics; and the International Center for Agricultural Research in the Dry Areas (ICARDA), Aleppo, Syria, for the arid regions. In addition to the centers, CGIAR supports associated institutes, two of which perform functions directly relevant to sustainable resource management. These are the International Board for Soil Research and Management (IBSRAM), Bangkok, Thailand, and the International Council for Research in Agroforestry (ICRAF), Nairobi, Kenya. Although there are many other institutions and organizations active in these regions, these international centers and institutes have become focal points.

Humid Tropics. Lal (*23*) listed the research and development priorities for the tropics. The first priority was to identify the reasons for the nonacceptability of the research information already available. Although the innovative concepts and subsystems discussed are technically sound, their economic evaluation and social acceptability need to be assessed for various socioeconomic environments. Technical concepts must be evaluated as integral parts of the overall system rather than as individual components of the improved technology. Soil scientists must become more active in working with economists and social scientists. Improved soil management is sustainable only within economically improved farming systems.

Management of soil structure is perhaps the largest problem in the tropics. The reasons that the structure of most tropical soils deteriorates so rapidly and how this can be prevented are still largely a mystery. Soil compaction and crusting are closely related problems, and additional research is critically needed. Lal (*23*) emphasized that a specific research priority for soil resource management in the 21st century is to acquire basic data from well-designed, long-term field experiments on soil management involving technology with various levels of inputs. A point clearly overlooked by many scientists is the importance of livestock in the tropics. Integrating perennial crops and livestock with food crop production is a high priority, especially in regions where rapid soil degradation is a severe problem with intensive arable land use.

Land clearing and development in forested regions are major problems. With increasing labor constraints, mechanized methods will become dominant. Unless a better understanding is gained about the effect of various clearing methods on various soil types, serious and sometimes irreversible degradation of the soil resource base may occur.

Much land is already seriously eroded and degraded, and methods should be developed to regenerate these acres. Knowing the critical limits of key soil properties for various levels of management is crucial in this endeavor. The development of appropriate tillage systems, which in many cases may include no-till or reduced till, is perhaps the greatest single need because tillage can be both very beneficial and destructive, depending on the particular soil, cropping system, and climate.

Semiarid Regions. The semiarid tropics are generally defined as the region within the tropics where the mean monthly rainfall exceeds mean potential evapotranspiration during 2 to 7 months of the year. Consequently, the most critical characteristic of the semiarid tropics is the definite seasonality of rainfall. As previously discussed, El-Swaify and associates (*10*) reviewed soil management for optimizing productivity under rainfed conditions in these regions. The most abundant soils in the semiarid tropics are Alfisols, and these soils are extremely vulnerable to erosion, crusting, compaction, drought, and limited rooting depth. Alfisols contain predominantly low-activity clays and have low plant-available water reserves. Improved management systems for conventional cropping of Alfisols have succeeded in increasing yields of conventional crops, largely due to improved cultivars and use of fertilizers (*10*). Effective practices for improving soil and water conservation, however, have not been developed. This is primarily because of the extreme structural instability of these soils. Therefore, a critical research need in the semiarid tropics continues to be the development of management systems that can sustain the soil resource base. This is indeed a challenge, as discussed previously in relation to figures 1 and 2. Alfisols are inherently low in soil organic matter, even under native vegetation, and once they are tilled, the organic matter becomes critically low.

Clearly, there has not been enough attention given to the livestock component in the semiarid tropics. Livestock is extremely important in these regions and in years of lower-than-anticipated precipitation animals may be used to consume as forage the crops that have insufficient water to mature. However, livestock can also lead to severe soil degradation if it is allowed to consume all of the vegetation and crop residues, which will lead to further decline of already low soil organic-matter levels. In soils with low-activity clays, as is the case of Alfisols, organic matter is essential to maintain base exchange capacity and soil structure.

FAO (*12*) used the growing period as the basis for assessing climatic resources in developing countries. The growing period is the number of days during a year in which precipitation exceeds half the potential evapotranspiration, in addition to a period required to use an assumed 100 millimeters (4 inches) of water from excess precipitation (or less, if not available) stored in the soil profile. Areas with a growing period between 1 and 74 days are classified as arid, and areas with growing periods between 75 and 119 days are considered semiarid. This classification scheme can be useful in assessing agricultural regions. However, some dryland farming regions, such as the southern Great Plains of the United States, have zero growing days by this classification because the half-potential evapotranspiration for every month of the year exceeds the average precipitation for the month.

The highest research priority in these regions, particularly in developing countries, is the development of soil and water management practices that will result in increased productivity over a relatively short term with few purchased inputs. Examples of possible technologies include contour terracing, tied ridges, water harvesting, fertilizers, and improved weed control. On-farm demonstrations and adaptive research need to be strengthened.

Arid Regions. Research priorities in arid regions, except where irrigation is practiced, must concentrate on livestock systems. The greatest concern in many of these regions is degradation of the soil resource by overgrazing. Degradation of the soil in these areas is often irreversible because the low precipitation and high temperatures makes it extremely difficult to regenerate the soils. Most often, severe degradation of these soils leads to desertification of the region. Scientists must work with policymakers and governments to develop the databases required to make sound projections, and governments must implement policies and programs that assure the soil resource base can be sustained.

In Summary

A larger population, higher incomes, and increased consumption of animal proteins all intensify pressure on soil and water resources. During the past several decades, technological advances have led to significant increases in crop production. At the same time, there is increasing evidence and concern that some of these technologies have triggered environmental degradation. Many of the technologies are also energy-intensive. Concern is mounting that agriculture production systems in many areas cannot be sustained.

The most important factors determining a sustainable system are climate and soil. Careful analysis of these factors can reveal a wealth of information for formulating guidelines and policies. The interactions of resources, technology, and policy environment require methodological approaches to (1) quantify how soil and climate resources determine potential technologies; (2) determine what impacts selected technologies have on soil resources; and (3) develop a strategy for selecting appropriate production technologies that meet economic objectives and environmental goals.

In addition to the emphasis on systematic land use planning and improved resource management, there is a need for monitoring soil changes as an integral part of agricultural development strategies. This is particularly true when dealing with marginal land that is readily vulnerable to degradation. Multidisciplinary teams should devise means for documenting changes in soil productivity, soil quality, farmer profits and preferences, and other measures of "sustainability." Such information would serve as a prognosis for early identification of problems and allow application of timely corrective interventions in order to protect national investments in agricultural development. Government policies are needed for both corrective and preventive land use planning. Preventive land use planning, the ultimate goal, would consider the inherent attributes and limitations of any site (soil, climate, and topography) for agricultural development and prescribe appropriate land uses for sustainable agriculture at the site, accordingly.

This chapter has focused largely on the dryland agriculture. It can be argued that dryland agriculture is more complex and heterogenous than irrigated agriculture or agriculture in more favored rainfall regions. This complexity and diversity spans land characteristics, preferred crops, environments, and people. A second green revolution is not likely until investments and commitments are made to bring the understanding and state of the art in these areas to a status comparable to those that resulted in the first revolution, which occurred in the irrigated and favorable rainfall areas.

Soil scientists cannot work alone or afford to engage in traditional research to meet these challenges. Rather, we must increasingly work with scientists from other disciplines to devise productive and sustainable systems, and we must do so creatively and efficiently.

REFERENCES CITED

1. Brown, L. R. 1987. *Analyzing the demographic trip.* In *State of the World.* W. W. Norton & Company, New York, New York. pp. 20-37.

2. Brown, L. R., and S. Postel. 1987. *Thresholds of change.* In *State of the World.* W. W. Norton & Company, New York, New York. pp. 1-19.
3. Buringh, P., H.D.J. Van Heemst, and G. J. Staring. 1976. *Computations of the absolute maximum food production of the world.* African University, Wageningen, The Netherlands.
4. Council for Agricultural Science and Technology. 1988. *Effective use of water in irrigated agriculture.* Report No. 113. Ames, Iowa.
5. De Vries, P.F.W.T., and C. T. De Wit. 1983. *Identifying technological potential for food production and accelerating progress in productivity.* In *Accelerating Agricultural Growth in Sub-Saharan Africa.* IFPRI, Harare, Zimbabwe.
6. Dregne, H. E. 1989. *Desertification of drylands.* In *Proceedings, International Conference on Dryland Farming.* Texas Agricultural Experiment Station, College Station.
7. El-Swaify, S. A. 1987. *Contrasting the conservation and management requirements of different soils for improved rainfed farming in the semiarid tropics.* In I. Pla Sentis [editor] *Soil Erosion and Productivity.* Venezuelan Society of Soil Science, Maracay. pp. 935-953.
8. El-Swaify, S. A. and E. W. Dangler. 1982. *Rainfall erosion in the tropics: A state-of-the-art.* Special Publication No. 43. Soil Science Society of America, Madison, Wisconsin. pp. 1-25.
9. El-Swaify, S. A., E. W. Dangler, and C. L. Armstrong. 1982. *Soil erosion by water in the tropics.* Research Extension Series 024. Institute of Tropical Agriculture and Human Research, University of Hawaii, Honolulu.
10. El-Swaify, S. A., P. Pathak, T. J. Rego, and S. Singh. 1985. *Soil management for optimized productivity under rainfed conditions in the semiarid tropics.* Advanced Soil Science 1:164.
11. El-Swaify, S. A., A. R. David, S. Sinanuwong, A. David, and A. Tengah. 1977. *Potential for saline water irrigation of tropical soils.* In *Proceedings, International Conference on Management of Saline Water Irrigation: Planning for the Future.* International Society of Soil Science, Wannigen, The Netherlands. pp. 358-375.
12. Food and Agriculture Organization, United Nations. 1978. *Report on the agro-ecological zones project: Methodology and results for Africa.* World Soil Research Report 481. Rome, Italy.
13. Food and Agriculture Organization, United Nations. 1981. *Agriculture: Toward 2000.* Rome, Italy.
14. Food and Agriculture Organization, United Nations. 1986. *African agriculture: The next 25 years.* Main Report. Rome, Italy.
15. Food and Agriculture Organization, United Nations. 1987. *Improving productivity of dryland areas.* Rome, Italy.
16. Haas, H. J., C. E. Evans, and E. F. Miles. 1957. *Nitrogen and carbon changes in great plains soils as influenced by cropping and soil treatments.* Technical Bulletin 1164. U.S. Department of Agriculture, Washingtn, D.C.
17. Hartmans, E. H. 1983. *African food production: Research against time.* Outlook Agriculture 12:165-171.
18. Higgins, G. 1984. *African resources of terrain, land and water for the populations of the future—management for development.* In *Advancing Agriculture Production in Africa.* CAB International, Wallingford, England.
19. Hornick, S. B., and J. F. Parr. 1987. *Restoring the productivity of marginal soils with organic amendments.* American Journal of Alternative Agriculture 2: 64-68.
20. Houghton, R. A., and G. M. Woodwell. 1989. *Global climate change.* Science of America 260: 36-44.
21. Keen, M. 1988. *Clearer thoughts flow on irrigation.* CERES (The FAO Review) 21(3): 33-35.
22. Lal, R. 1982. *No-till farming.* Monograph 2. International Institute of Tropical Agriculture, Ibadan, Nigeria. 68 pp.
23. Lal, R. 1986. *Soil surface management in the tropics for intensive land use and high and sustained production.* Advances in Soil Science 5: 1-109.
24. Murthy, R. S., J. C. Bhattoucharjee, R. J. Landey, and R. M. Pofali. 1982. *Distribution characteristics and classification of vertisols. Vertisols and rice soils of the tropics.* In *Managing Soil Resources.*Proceedings, Twelfth International Congress of Soil Science. Indian Society of Soil Science, New Delhi, India. pp. 1-27.
25. Norse, D. 1976. *Development strategies and the world food problem.* Journal of Agricultural Economics 27: 137-158.
26. Parr, J. F., R. I. Papendick, S. B. Hornick, and R. E. Meyer. 1989. *Strategies for developing low-input sustainable farming systems for rainfed agriculture.* In *Proceedings, International Symposium for Managing Sandy Soils.*Central Arid Zone Research Institute, Jodhpur, India.
27. Pierce, F. J., W. E. Larson, R. H. Dowdy, and W.A.P. Graham. 1983. *Productivity of soils: Assessing long-term changes due to erosion.* Journal of Soil and Water Conservation 38: 39-44.
28. Rangeley, W. R. 1985. *Irrigation and drainage in the world.* In W. R. Jordan [editor] *Proceedings, Water and Water Policy in World Food Supplies.* Texas A&M University, College Station.
29. Rodale, R. 1988. *Agricultural systems: The importance of sustainability.* Phi Kappa Phi Journal 2: 6.
30. Roose, E. J. 1977. *Adaptation of soil conservation techniques to the ecological and socioeconomic conditions in West Africa.* L'Agron. Tropical 32: 132.
31. Ruttan, V. W. 1989. *Sustainability is not enough.* Better Crops with Plant Food. (spring): 6-9.
32. Soil Survey Staff. 1975. *Soil taxonomy. A basic system of classification for making and interpreting soil surveys.* Handbook 436. U.S. Department of Agriculture, Soil Conservation Service, Washington, D.C.
33. Steiner, J. L., J. C. Day, R. I. Papendick, R. E. Meyer, and A. R. Bertrand. 1988. *Improving and sustaining productivity in dryland regions of developing countries.* Advances in Soil Science 8: 79-122.
34. Swindale, L. D. 1988. *Agricultural development and the environment: A point of view.* In *1987/1988 Annual Report.* Consultative Group on International Agricultural Research, Washington, D.C. pp. 7-14.
35. The World Bank. 1986. *The World Bank Atlas.* Washington, D.C.

12

Natural Resources Assessment and Policy

Charles M. Benbrook

Government programs and policies have always had a major influence on the use of natural resources. Some policies strive to encourage resource development by providing low-cost capital, technical assistance, and access to or establishment of infrastructure (roads, canals, and ports). Other policies evolve in response to problems in the way resources are used: conservation and environmental policies deal with the physical and biological consequences of resource use; zoning, taxation, and other policies have impacts on land use; and a variety of other policies deal with the control and ownership of natural resources.

New policies are adopted or old policies are changed in democratic societies when a consensus emerges that a different balance is needed in the fundamental objectives or outcomes sought or sanctioned through government programs or policies, or when a consensus emerges that there is a better way to attain some desirable objective.

Natural resource conservation policy is in the midst of what may come to be viewed as its second major phase of evolution in this century. The first sea-change occurred in the decade following the Dust Bowl. It culminated in the mid-1940s with the establishment of the basic public policies, approaches, and institutions that still form the basis for delivering to land operators technical and/or financial assistance in developing or conserving natural resources.

The second wave of major change in natural resource use and conservation policy began in 1984 or 1985. It was set in motion in the early 1980s by environmental concerns arising from agricultural and forest-product industry operations. The debate and consensus-forming process leading to the 1985 farm bill's historic Conservation Title was an important watershed and will likely be remembered as the first concrete step in defining a new direction and approach to resource use policy.

This new period of policy evolution is far from over. My sense is that three major additional milestones must be reached before this current phase of change in natural resource management policy will be complete.

Hopefully we will recognize and thoughtfully respond to the interconnectedness of the resource base itself and the many complex ways in which agrichemicals, water, sediment, and cropping systems interact. The laws of nature cannot be amended and should not be ignored in evaluating and refining policy.

It is already evident that current policies need to foster a more balanced and systematic approach in helping land managers control erosion, protect water quality (both surface and groundwater), and enhance wildlife habitat. Also, recognizing the interconnectedness of resources is a necessary step in establishing a firmer biological and ecological basis for determining how, where, and when resource use and conservation practices should be altered. Without passing this milestone, we may continue to create new problems each time we initiate new steps to solve old problems. This capacity to more firmly appreciate the biological interconnectedness of agroecosystems will become particularly crucial in areas where pest pressures and pesticide use add a public health/wildlife impact dimension to the equation.

Today, land use is treated as a given by conservation policy and resource management specialists; in the future, land use needs to be viewed as a critical point of intervention, at least in a growing range of circumstances and areas.

Traditional conservation plans and other resource stewardship evaluations generally do not begin with the question "Is this land being used properly in light of its inherent physical potential?" As a result, conservationists have never really considered the benefits of a change in land use as a policy-driven remedial intervention, unless a change in land use is otherwise consistent with economic opportunity or government programs like the Conservation Reserve Program (CRP). As a society, though, we are reaching a point where the landowner's right to pursue any land use, regardless of the social costs and consequences, is outstripping conservation technologies and the public's willingness to pay for remedial actions (or accept environmental degradation). Indeed, in some parts of the country, truly heroic, but probably ill-fated, efforts are now being made in trying to develop affordable conservation systems that can sustain current land use patterns in regions where serious problems are now evident (erosion, groundwater contamination, and pest control problems).

We also must pay much closer attention to the costs of conservation and environmental protection efforts. These costs are real and growing. High costs may be justified in certain regions in the event of higher real crop and commodity prices. But we must be realistic and recognize that it is more likely that competitive pressures in the next decade or so will keep crop prices from rising much above current levels. Accordingly, conservation practices and systems will have to make economic sense in a world characterized by narrow profit margins and careful cost accounting. That is why I am convinced that economic forces will eventually prevail over current policies that lock in misguided land use patterns in some regions and that cropping patterns in areas now experiencing both economic and environmental problems will finally begin adjusting in ways that markedly reduce resource problems and bring back some realistic prospect of sustainable, profitable production. New policies, like the CRP, may evolve to accelerate this healthy trend.

The idling of productive land resources through government programs will, finally, pass into the history books.

The idling of productive land makes no sense. It is wasteful and an excessively expensive, often counterproductive way to help support farm income—a desirable goal—by reducing surplus crop production. As a nation, we simply can no longer afford to continue routine use of such a misguided policy tool. As routinely practiced in the context of current farm programs, the idling of land does little to build soil structure or fertility. It imposes unnecessary and increasingly burdensome costs on farmers who must pay taxes, mortgages, and rent and must control weeds on idled acres; it artifically and unconstructively inflates the value of land; and it encourages more intensive, often inappropriate, and environmentally damaging uses of other land.

The policy process and fiscal reality has begun to wean American agriculture from its infatuation with land retirement as the only feasible method to balance the supply and demand for major commodities. The so-called 50-92 provision of the 1985 farm bill is a step in the right direction; this process was advanced conceptually in the discussion of decoupling as a policy alternative in the last few years, and may well come to full fruition in the 1990 farm bill. The new farm bill's commodity program provisions are likely to continue the evolution in policy toward several major changes:

► A greater degree of flexibility will be allowed in how farmers use crop acreage bases.

► Deficiency payments per farm will be based on a somewhat different mix of factors, and will be limited in size to a given farmer or for a given field on some basis other than imposing a 15 to 20 or 25 percent set-aside requirement.

► Paid land diversion programs will be phased out as routine and, by the mid-1990s, even occasional policy tools.

► Economic use for haying, grazing, and trees will be allowed, indeed encouraged, on land enrolled in the CRP or in other long-term programs designed to bring about more stable, resource-conserving land use patterns.

Production adjustment goals will be met by altering the acreage enrolled in the CRP and other environmentally driven land reserves, by encouraging or requiring more diversified crop rotations, by raising the costs or otherwise discouraging use of certain inputs and practices that cause water quality problems, and by lowering target prices.

The Emerging Challenge

Future natural resource policy will be shaped by contemporary challenges, whatever they may be and however they evolve in the public's eye. Clearly, in the 1990s, environmental and water quality concerns will be major factors shaping policy. Beyond these new issues, there may also be an important change in attitude. In the past, the principal goal of conservation, embedded in current policies, has been to preserve resources in their natural, productive state, subject to the activities of man. In the years ahead, pressures on the land will surely grow more intense. In most agricultural regions, farmers will have to increase, not just sustain, the productive capacity of soils, and they will also have to increase the quantity of food produced on each acre of crop or rangeland. Conservation policies will be needed to help farmers achieve and sustain higher levels of production. More intensive land use patterns will evolve, often including double-cropping and rotations. These challenges will be driven both by markets and policy, and they must be accomplished at the same time that farmers figure out how to reduce adverse impacts on the environment.

Economic pressures on farmers are also likely to grow, driven by the need to remain competitive and contribute more fully to national economic welfare. Because of fiscal limitations, government programs are, in general, not likely to assume a greater share of the cost of conservation investments. Accordingly, conservation and environmental protection efforts will need to be largely self-supporting, and provide land managers tangible benefits through gains in the efficiency of cropping systems, particularly in the utilization of purchased inputs.

As the debate unfolds in the coming years on natural resources policy, the costs and benefits likely to arise from alternative courses of action will be a recurrent matter of analysis, debate, and sometimes contention. Consensus may not be any easier to reach, but the dialogue will be greatly enriched by the work of William Larson. Dr. Larson has pioneered many key applications of science to the national assessment of natural resource management issues and policy challenges.

The task assigned to me—summarizing Dr. Larson's contributions to the policy process—is a difficult one. I have chosen to approach it unconventionally. Throughout his career, he has found ways to explain the need for resource conservation to many people outside his discipline, indeed, many people not involved with research or agriculture. His ideas, research results, and policy insights have mattered because he has taken the time to communicate broadly with people.

Dr. Larson has also demonstrated the value of multidisciplinary collaboration. He is a problem-solver. He thrives on figuring out a viable way to address technically complex and multifaceted problems that may seem, on the surface, both simple and practical. Examples of the questions Dr. Larson has addressed in his research include the following: How to identify highly erodible land? What role can residues play in erosion control? Would it matter to agricultural sustainability if 10 percent of the nation's petroleum needs were met by ethanol from crop residues? Can sewage sludges be recycled on farmland without long-run problems?

As the resource conservation policy debate evolves in the 1990s, a number of issues are bound to surface: enforcement of the 1985 farm bill's conservation title, approaches to nitrate-water quality problems, the biological basis of alternative farming systems, the integration water quality protection goals into ongoing soil and water conservation policy and commodity policy. During his research career, Dr. Larson advanced new scientific approaches that are now being used to evaluate these issues. In many instances, the results of his work have played a key role in shaping and administering new policies. But what about emerging issues and policies? The following discussion was taped on April 10, 1989, at Dr. Larson's home in St. Paul, Minnesota. His thoughts represent a uniquely well-informed view of evolving resource conservation issues and a record of the remarkable intellectual breadth of this man's many contributions.

The Interview

BENBROOK: *In your estimation, what are the two or three most pressing natural resource management challenges facing farmers in North America today?*

LARSON: We should think about better use of our land, not just our farmland, but also our forest land, recreational land—all land. We in America have thought of our land as inexhaustible, something that did not need a lot of attention because we had plenty. We know now that attitude is not right. We have a lot of land that is misused. We don't have too much land; we simply have too much land in the wrong use. We are starting to take steps to correct that, but it is still a big problem.

Brief Review of Dr. Larson's Career

BENBROOK: *After finishing your doctoral dissertation on substitution of sodium for potassium in 1949 at Iowa State University, you went right into an assistant professorship at Iowa State?*

LARSON: Yes, I stayed on at Iowa State a short time after I received my degree, largely working in the field of soil fertility. But I had decided that I needed some experience working elsewhere. I wanted to get away from the Iowa-Nebraska area where I grew up and where essentially all my experience had been. So, in 1950, I joined the Agricultural Research Service (ARS) and worked at Montana State University in Bozeman.

I worked on irrigated agriculture problems in the Yellowstone River Valley. At that time, one of the big problems was irrigation efficiency. The major crop was sugar beets. Farmers tended to irrigate far too heavily. As a result, the soils were too wet, and compaction was a problem. That research experience is what got me interested in the physical properties of soils.

BENBROOK: *The Bureau of Plant Industry, was that a predecessor of ARS?*

LARSON: Yes, I think it was in 1952 that the Bureau of Plant Industry was combined into the ARS. At that time the soil survey responsibilities of the old Bureau of Plant Industry were transferred to the Soil Conservation Service (SCS). In turn, the ongoing research activities within SCS were transferred to the new ARS.

BENBROOK: *Then you went back to Ames with ARS in 1954?*

LARSON: I stayed with ARS and moved back to Ames to join the tillage conservation team at Iowa State University. At that time there was a lot of interest just starting to develop on better conservation tillage practices—the mulch tillage or residue-type tillage practices that we know today. They had been developed further in the Great Plains and were just beginning to receive attention in the Corn Belt. There was a lot of talk about minimum tillage, such things as plow-plant tillage, wheel-track planting, and so forth. So many of the forerunners of our current conservation tillage systems were developed in the late 1950s and 1960s.

BENBROOK: *You continued to work at*

Ames through 1967?

LARSON: Yes. In 1965-1966, I spent a year in Australia as part of an exchange program between that country's federal agency (CSIRO) and our federal agency (ARS). In Australia, I learned some new soil mechanics theories. Scientists in Australia were more advanced on soil compaction in the mid-1960s than we were in this country. At that time, the center of knowledge in terms of soil structure, soil compaction, and basic theory of soil mechanics was in Adelaide, Australia. When I came back, ARS asked me to move to the University of Minnesota to head up a soil structure unit. This fit well with what I had learned in Australia, so in 1967 we did move to St. Paul.

BENBROOK: *In 1967, then, you became the national technical leader in tillage and residue management with ARS, and you received a concurrent appointment as professor of soil science at the University of Minnesota. That's when your dual career with ARS and academe began?*

LARSON: I really had the same kind of appointment at Iowa State, too, but I became more active in graduate training after I moved to Minnesota. I also began assuming various national responsibilities in ARS, in terms of program planning and priorities.

BENBROOK: *Then in 1982 you left ARS and became a professor and head of the Soil Science Department at the University of Minnesota?*

LARSON: In 1982, Dr. Martin, the head of the Soil Science Department at Minnesota, retired. I thought it would be a new challenge to join the university and be head of the department. I had been located in the department since 1967. It was a new challenge, and I always liked challenges, so I applied for the job and they gave it to me. I have enjoyed it very much.

Conservation Provisions of the 1985 Food Security Act

BENBROOK: *Let's talk a little bit about the conservation provisions of the 1985 farm bill. Many aspects of your work over the years made an important contribution in developing the analytical tools and knowledge that made it possible to both develop those provisions and make the case for them. Do you think the conservation provisions of the 1985 farm bill—the sodbuster, swampbuster, conservation compliance, and the Conservation Reserve Program—will make a substantial difference in the nation's effort to control soil erosion?*

LARSON: The CRP was certainly a step forward. We had so-called conservation reserve programs before, but the CRP had more teeth in it. The CRP had provisions in it so that highly erodible land was the only land that was eligible, rather than any land a farmer wanted to put in. Since passage of the act, there has been a lot of debate over how to identify "highly erodible" land to bring into the CRP and in what order of priority? Some of us have felt that it ought to be done on a sound scientific basis that uses detailed data bases. We can now make good use of scientific principles in analytically approaching the task of determining which land is highly erodible and which acres will be damaged. We have tried to take the national data bases and show that you can analyze these data and come up with recommendations that are feasible and guidelines that identify that land that should be taken out of crop production first and put into another, more stable land use.

BENBROOK: *What about the other provisions? Will they make a big difference over the long run?*

LARSON: Well, I know less about them. I think the swampbuster and sodbuster programs are good programs and are developing fairly well. The cross-compliance provision is more debatable. The general philosophies underlying these provisions are fine, but their effectiveness will depend on how they are administered and enforced.

Importance of New Data Sources and Analytical Innovations

BENBROOK: *How critical were the 1977 and 1982 National Resources Inventories (NRI) in the development and passage of the conservation provisions in the farm bill?*

LARSON: The 1977 and 1982 NRIs were

a giant step forward because, for the first time, they gave us an adequate, or at least semi-adequate, inventory of the nation's soil and water resources. Analysis of NRI data demonstrated the value in taking these national inventories and using them in a technically sound way to administer programs, identify erodible land, gauge the extent of conservation needs, and more cost-effectively administer conservation programs. We have not solved all of the problems, but I think we now know how to approach the problem. The big remaining problem is ironing out all the technical details, and getting more complete data on soil properties, land use, and resource needs.

BENBROOK: *Do you remember where and when use of the physical factors of the universal soil loss equation (USLE) as a measure of inherent erodibility began?*

LARSON: I think I first became aware of that new analytical method by working with you, Chuck, and with Arnie Miller, back in 1980 or 1981. I think that was a great step forward.

BENBROOK: *Did you ever wonder why two economists developed this approach to analyzing soils data?*

LARSON: If you look at science, often times people a little bit outside of the mainstream come up with new ideas, whereas people who work most of the time within the discipline get ingrained in a way of approaching problems and never think about new ways to analyze things. I think that you will find this to be true in many areas of science—people a little bit outside the field often come up with important new ideas.

BENBROOK: *When you first began working with the physical factors of the USLE (the RKLS factors) separate from the cropping and management factors (the CP factors), what were the most intriguing and important applications you thought could be made of this analytical approach?*

LARSON: We used the USLE with available data bases (NRI, SOILS 5) to identify land that ought to be taken out of the usual crop production patterns because of high erosion losses. We thought the approach could be used to assess how much damage was actually occurring from erosion. Back in the 1970s, we used a similar approach for other things, including a national study on how much crop residue could be removed for energy production without exacerbating soil erosion.

BENBROOK: *When did you decide to develop the soil erosion-soil productivity prediction model that you have done so much with in recent years?*

LARSON: I suppose it was about 1979. At that time I was still with ARS, and I was called frequently to travel to Beltsville, Maryland, to help with the Resources Conservation Act (RCA) assessment. I recall vividly that at one frustrating meeting no one seemed to know how to express erosion in terms of potential costs or in terms of losses in production. I came back and suggested to one of my graduate students, Fran Pierce, that we could develop a simple model to get at this problem. Fran Pierce and I then tried various simple models. We were not happy with any of them, but we thought we were making progress. About that time we became aware that a graduate student, Laurie Kiniry, at the University of Missouri, along with her major professor, Dr. Clarence Scrivner, had a model that was very similar to what we had been working on. So we called Dr. Scrivner and asked if we could further develop the basic concepts they had incorporated into their model.

I was excited at that time about the use of computers and national databases in making national assessments of conservation needs and opportunities. So, Fran Pierce, Bob Dowdy, myself, and others applied our model to Minnesota and Iowa soils; these were the areas we knew best. Everything seemed to check out reasonably well. You will recall that in developing the erosion-productivity model, we made some assumptions to help bound the analysis. We assumed that there are two kinds of attributes of soil: replaceable ones and nonreplaceable ones. We focused our analysis only on the nonreplaceable ones because these seemed most important in the context of the RCA process. Subsequently, we expanded the application of the model throughout the Corn Belt and published our work in various journals. I think it has been received rather well.

Our conclusions have agreed with the results from other, often more complex models.

We have also tried the model outside of the Corn Belt, which required us to change it some and add new factors. We had an international study with Dr. "Reds" Wolman and Frank Reijberman from The Netherlands and other studies that were published in the *Journal of Soil and Water Conservation*. We have learned from these applications that you have to compromise sometimes between detail in models and the availability of data to use in applying the model. A model built to run on nonexistent data or data that can be compiled only at considerable cost for a limited number of soils or regions is of limited use in addressing national resource management needs.

The Technical Basis for Conservation Compliance

BENBROOK: *I would like to return to the conservation compliance provision of the Food Security Act. This provision requires farmers with highly erodible land to come in to their local SCS office and work with the SCS staff to develop a conservation plan that would bring erosion down substantially in order to maintain future eligibility for commodity price-support program benefits. The policy's goal is to bring erosion down, if not to the T value, then reasonably close to T, using some combination of tillage, residue management, contour planting, crop rotations, terraces, and other conservation practices. In the last two years some one million plans have been developed, and, beginning next year, farmers will start the process between 1991 and 1994 of getting those plans onto the land. That was the theory and the concept and the goal of conservation compliance. How well do you think it is going to work in practice out on the land?*

LARSON: I think it can work and should work. However, administration of the provision is going to have to be tough.

BENBROOK: *How important will conservation tillage and residue management be in farmers' plans?*

LARSON: Conservation tillage and residue management is the basis of most of our plans. It is the one tool that we have that has been developed in the last 10 or 20 years that is really a major step forward. It has got to be the major tool that we use in most compliance plans.

BENBROOK: *Most compliance plans in the Midwest are calling for at least 30 percent soil cover to meet the erosion control goal. Is that degree of soil cover common today, and do you think farmers will have an easy time achieving routinely that level of cover?*

LARSON: No, 30 percent cover is not common today, even in the Corn Belt where most farmers practice some form of conservation tillage. Several analyses have tended to overestimate how much residue is actually left on the land. It is rather difficult to measure the amount of residue on farms in an area. But I do think the extent of soil cover has often been exaggerated. Most of the estimates in erosion surveys are based largely on crop history and on what kind of tillage tools have been used. But tillage tools can vary. You can use a chisel plow and bury all of the residue if you want, or you can use a chisel plow and leave most of it on the surface, depending on how you adjust the tool and use it. Several actual field studies where soil cover has been measured have found this to be the case.

BENBROOK: *Bill, let's walk through some common fall tillage practices on a corn field harvested for grain. Give me a ballpark estimate of the amount of cover that would be left on the surface following these combinations of tillage practices. We'll start from what might be called conventional tillage practices and work toward higher degrees of conservation. First, a farmer fall plows the field, spring disks it, and prepares it for planting. What kind of cover would be left in the spring in that kind of a system?*

LARSON: Essentially none.

BENBROOK: *Now, assume fall tillage involved only a light disking or chisel plowing, followed in the spring by another pass with the disk and perhaps a field cultivator. How much residue might be left in this case?*

LARSON: Following a 100-bushel crop of corn, a light disking and field cultivator would probably leave about 30 percent soil cover.

BENBROOK: *So a farmer could make the basic 30 percent soil cover goal in many compliance plans with a light disking?*

LARSON: Yes, assuming the farmer had the residue from 100 bushels or more of corn to start with.

BENBROOK: *Is there a simple rule of thumb about the percent reduction in residue cover expected following different tillage practices? Or are such ballpark estimates impossible because so much depends on how tillage tools are adjusted and operated?*

LARSON: Soil cover following use of a chisel plow, for example, depends on what kind of a shank you use, how deep you set it, ground speed, soil moisture conditions, and many things. So the amount of cover following the use of any common tillage tool can vary a great deal. There are some general estimates in the literature.

BENBROOK: *What about the use of a heavy disk? If a heavy disk is used, isn't the heavy disk going to bury at least half the available residue in a given pass?*

LARSON: Yes, a heavy disk will bury perhaps half the residue in one pass. If the soil is relatively dry and the residue is relatively dry, it probably will leave a lot of the residue in the top inch or two of soil. There is still some debate about how effective such buried residue is in controlling erosion. It is probably somewhat effective, but not as effective as residue retained on the surface.

BENBROOK: *The alternative conservation systems (ACS) policy adopted by SCS involved the relaxation of conservation compliance erosion control goals because of perceived adverse economic impacts on farmers. The ACSs incorporated in compliance plans are designed to assure a substantial reduction in erosion from prepractice rates, but not necessarily to T. Was this policy needed?*

LARSON: My rule of thumb is that with good tillage practices, even when a farmer is following an erosion-prone rotation, like corn followed by soybeans, we can now profitably reach T with well-designed conservation tillage systems on anything up to an 8 percent slope. Over 8 percent slope it gets pretty tough to reach T with tillage practices alone.

BENBROOK: *Isn't 8 percent slope, though, pretty steep?*

LARSON: Eight percent slope is pretty steep. I have forgotten the exact figures, but of the cultivated land in the Corn Belt, there might be a little over 5 percent of the land that has a slope greater than 8 percent. So we are really talking about relatively few acres—much of it already in the CRP.

BENBROOK: *A major concern in SCS and ASCS [Agricultural Stabilization and Conservation Service] field offices is who is going to be responsible for enforcing a farmer's adherence with the practices called for in a conservation compliance plan. As you travel around the state, what ideas are you picking up in the field regarding who should be responsible for enforcing compliance?*

LARSON: Well, if history is a guide, SCS would check compliance, and ASCS would administer programs subject to compliance. I suppose this might prove an awkward way to handle things, having two different agencies involved. Beyond that, I really don't know.

Nitrogen Use and Management

BENBROOK: *Let's talk about nitrogen management. Why do we still lack a reliable method to measure the available nitrogen in agricultural soils?*

LARSON: Because it really is a rather difficult analytical task. In a good Corn Belt soil, there is perhaps 10,000 pounds of nitrogen in the top foot of soil over each acre. That is a lot of nitrogen. There are many, many different kinds of organic compounds in the soil that interact with nitrogen and affect its availability. And when we measure the 10,000 pounds, we measure them all, regardless of their degree of availability. Some of it is very resistant to decomposition; some of it is very readily decomposed. So that is one problem. The other problem is that mineralization of the nitrogen, converting it from organic matter nitrogen to mineralized forms that are available to plants, and available for leaching, is a biological process that requires moisture, heat, and oxygen. Microorganisms that do the job are very sensitive to temperature, water, and the oxygen content of soil.

BENBROOK: *Do they also need a source of energy?*

LARSON: The organic matter is a source of energy. The rate of mineralization is really difficult to predict, just like the weather. Indeed, it is really a weather-related phenomenon. It is very tough to predict what the moisture content, the oxygen content, and the temperature of the soil is going to be a year in advance. But if we could, we could estimate much more accurately the amount of nitrogen likely to become available in a given season. We have tried to fashion better measures by chemical means, measuring the fraction of the organic matter that is most susceptible to decomposition. I think there has been a fair amount of progress in this approach. And we have tried to develop models that rely upon the temperature-moisture-oxygen content of the soil. We have one here at Minnesota that works quite well, and if we knew exactly what the weather was going to be, we could probably predict soil-available nitrogen pretty accurately.

BENBROOK: *Do you think some of the recently developed tissue testing and related nitrogen measurement tools for calibrating the plant's need for nitrogen will evolve into practical and reliable field-level methods for making fertilizer recommendations?*

LARSON: I think they can evolve into reliable and usable methods, but they require some new technology. First of all, they require methods for taking an adequate and representative sample of plant tissues over a field. It is important to get the samples into a laboratory almost immediately, and then the laboratory needs to make the analysis quickly and cheaply so the results can get out within a couple of days. It can be done today, but we don't have the mechanisms we need to do it as well or as widely as we should.

The other thing we need to keep in mind is soil variability. Currently, we farm fields; we don't farm soils. But in a field you may have half a dozen different soils; in the extreme case, such as here in southern Minnesota, we may have deep, productive, silty clay loams and thin, coarse-textured soils on glacial knobs within the same field and be farming both soils within the field the same way. The yield potential of those two soils may be 80 bushels of corn in the case of the thin coarse-textured soil compared to 150 bushels for the silty clay loam soil. Yet we fertilize both soils the same; we use the same amount of pesticides and the same tillage practices across that field. We like to farm 80-acre fields here in Minnesota and Iowa, and the trend is toward larger, not smaller fields. The technology is already here to farm different soils within a field in different ways. For example, we can put a computer in the tractor with a soil map and pull equipment that can adjust fertilizer and herbicide rates, on the go, from one soil to another.

BENBROOK: *How do these machines know when they have gone from one soil type to the next?*

LARSON: Usually, a laser mounted on the machine directs beams toward a few targets positioned at field corners. The computer keeps track of where the machine is in the field by tracking these beams and also follows where the machine is on a computerized soil map.

BENBROOK: *So changes in input levels can be driven off a map of the soils within a field. Are there systems that can do this via satellite?*

LARSON: Yes, that's right. There are a number of things that we can do. We have modern soil survey maps now in most of the Corn Belt. These are being digitized so they can be put right in the cab of the tractor or the combine. We can take soil samples, feed the information into the soil map, and change our fertilizer rate depending on that soil test value. Another real possibility—in fact, it is being done now—is to put measuring equipment on the combine so that when we traverse a field we get a continuous readout of the yield. Then, next year, when the farmer goes over that same field, input levels could be calibrated more precisely in light of each soil's yield potential.

BENBROOK: *So in the 1960s and 1970s research focused on soil erosion controls. In the 1990s and in the first decade of the next century, we might follow through with a research program allowing us to develop a computer characterization of soil properties within a field, which can be further checked by monitoring actual yields, and, once a field is mapped, it would be mapped forever and such knowledge would make possible a whole new generation of management techniques, regardless of the cropping sequence used.*

LARSON: Right, exactly.

BENBROOK: *Such a research and development program would be an investment that would pay for itself many times over as the land continues to be farmed. The technology to use these information systems in application equipment would probably evolve pretty quickly. Do you feel that over a decade or so this technology will be refined enough to use widely and the cost will come down so that it would indeed be a worthwhile investment?*

LARSON: Yes, I do because then we can put the amount of nutrients or pesticides onto each part of a field in light of crop needs, but none in surplus.

BENBROOK: *In general, Bill, do you think we are overfertilizing the shallow, poorer soils within fields, or underfertilizing the richer parts of fields that have a higher yield potential?*

LARSON: I suspect we are overfertilizing the shallow soils and knobs.

BENBROOK: *What about established yield goals and the yields that most farmers strive for? Are they realistic in Minnesota or Iowa? How do yield goals and a soil's physical limitations interact in determining agriculture's contribution to water quality degradation?*

LARSON: The debate about yield goals is something people will argue about for a long time. There was a study in Nebraska showing that farmers' yield goals were often 20 percent too high. It was a statistically designed survey, so I presume it is right. I think yield goals probably are, in general, a little high. They are about right for the good years. They are too high for the bad years. Over time, they tend to increase, particularly after a few good years in a row. Every farmer's yield goal in Minnesota was probably too high last year because of the drought, but I suspect few will lower their future expectations.

BENBROOK: *Do you think there will be much change in application rates of nitrogen, phosphorus, and potassium in Minnesota this year, taking into account the likely carryover from last year's underutilized nutrients?*

LARSON: I have talked to farmers and they all say they want to take that into account. I think there will be some reduction, but I am not sure there will be enough reduction.

BENBROOK: *Or as much as there could be?*

LARSON: I recall vividly one farmer I talked to who said, "Well I am going to split the difference. Just to be safe, I am not going to put out as much as I did last year, but I am not going to cut back as much as some recommend." I suspect that is what is going to happen on most farms.

BENBROOK: *When making fertilizer recommendations to farmers, do extension agents, fertilizer dealers, and soil scientists pay adequate attention over the years to nitrogen and other nutrients that are carried over or come from other sources, such as legumes, manure, and the atmosphere?*

LARSON: I think the answer unquestionably is no, they do not.

BENBROOK: *That hasn't changed recently, or is it just beginning to change?*

LARSON: I think it is just beginning to change. Many extension programs are really emphasizing that point now. We do go through cycles. Prior to World War II, manures and alfalfa, sweet clover, and green manure crops were an important source of nitrogen, and we all recognized it. Then, after World War II, we got so hung up with fertilizers that we essentially told farmers that manure was not worth the cost of hauling it out of the barn. Our extension people, people that work in the field, now feel that manure management is an important problem here in Minnesota, particularly in the areas where there is a lot of alfalfa in rotations and a lot of manure applied to cropland. We are underutilizing manure as a nutrient resource, and we are not giving it proper credit. In southeastern Minnesota that might be the number one environmental problem we are facing.

BENBROOK: *Are you more concerned with nitrates or pesticides as a water quality problem?*

LARSON: I am more concerned with pesticides.

BENBROOK: *Why?*

LARSON: I think we can handle the nitrate problem. We have most of the technology that we need. We understand nitrogen behavior in soils reasonably well, and I think it is a matter of fine-tuning and education to take care of remaining management problems. I am not so sure about pesticides though. We don't know the chemistry of pesticides in soil as well as we know the chemistry of nitrogen. Nitrate is relatively short lived in the soil, but some pesticides are more persistent. They apparently can build up in the soils and may be there for a long time. We simply need to know more about potential pesticide problems.

BENBROOK: *Let me ask you about the behavior of nitrates in the soil. Once nitrates leach down below the root zone in a common midwestern soil, does the nitrate stay relatively stable until it eventually is washed down into either a shallow or deep aquifer?*

LARSON: Yes, once it is below the root zone it is stable and stays there, or moves deeper. There are no organisms to convert it. Secondly, because there is no organic matter below the root zone, there is no energy source for organisms to use. Once it gets below the root zone there is nothing that is going to change it.

BENBROOK: *How long does it take nitrogen to leach down below the root zone? How long does it take in a productive midwestern soil, not in a region typified by karst geography, to reach a shallow water table that might be 10 or 20 feet below the surface? Just a ballpark figure—2 to 5 years, 5 to 10 years?*

LARSON: I would say 2 to 5 years.

BENBROOK: *What about a moderately deep aquifer, 50 to 150 feet?*

LARSON: Ten to 20 years I would guess, depending on the rainfall, the nature of the vadose zone, and other factors.

BENBROOK: *Given what we know about the time period for nitrogen to leach down into underground aquifers, what do you feel we know now or can infer about the fate of pesticides moving downward through the soil? Can we use nitrate mobility and current levels of nitrate in groundwater as a basis to predict future trends and herbicide levels in groundwater?*

LARSON: That is out of my field of expertise. My impression would be, no, we cannot use knowledge about nitrate to predict pesticide contamination. The chemistry of pesticides in soil is extremely complex, and the types of pesticides that we apply vary greatly. We have some that are relatively soluble in water, some that tend to bind on the soil very tightly, and others that are broken down. So it is hard to generalize. Nitrate is a little easier because the chemistry is all the same.

Sustainable Agriculture

BENBROOK: *What is your concept of sustainable agriculture? What does it mean?*

LARSON: I think it means just what the term implies. It means a production system through which we can sustain our basic soil and water resources indefinitely and still produce food and fiber without endangering the environment.

BENBROOK: *Do you feel that sustainable agriculture will prove, as a concept, to be a passing fancy?*

LARSON: No, I think sustainable agriculture is not a passing fancy. We may change the terminology to describe it, but we would be less than responsible if we did not believe in sustaining our natural resources. There are a lot of things that need to be considered in a sustainable agricultural system. Proper land use, proper use of legumes in rotations, green manure crops, use of crop residues, manures, and other wastes are all important ingredients. There are a lot of other organic wastes, including municipal wastes, that probably can be better utilized. These systems are not simple, but it can be done. Again, land use is the key issue. We have lots of needs for land, and we have to spend a lot more time to ensure that each piece of land is used for the right purposes and managed in the right way.

BENBROOK: *For the purposes of discussion, could you consider and describe the essential differences in the biological processes going on within a productive midwestern clay loam soil under a corn-soybean-small grain-hay rotation, managed intensively for high yields where there is limited, but some, commercial fertilizer used, some manure applied, weed control is largely accomplished through cultivation, and pest problems are controlled through cultural practices, in contrast to a continuous corn or a continuous corn-soybean rotation with conventional tillage and conventional application levels of fertilizers, herbicides, and insecticides.*

LARSON: Well, that's a big order. In the first system where we are talking more about rotations and the use of organic materials, more organic materials will be added back to the soil. The microorganisms in the soil are probably much more abundant. They are turning this organic matter over continually, which, in turn, has a big influence on the structure of the soil, the physical condition of the soil.

BENBROOK: *How does soil structure change? Does it make the bulk density go up or down?*

LARSON: Generally, it makes bulk density go down. Soil particles are very small. A good Corn Belt soil typically will consist of 30 percent of clay. That means that 30 percent of the soil particles are less than two microns in size. But in most soils, when we pick up a handful, we see soil aggregates. These are small particles clustered together to form what we call aggregates. These aggregates are bound together by a number of things, including both the inorganic fraction of the soil and the organic fraction. These organic materials are in a constantly changing state, depending on the microorganisms present and their activities.

The first part, the chemistry of soil aggregation and structure, you cannot do much about. The mineralogy does not change; it is essentially constant over thousands of years. The organic matter fraction you can change. It is the part we have to work on. Some organic fractions are rather resistant and really do not change much over time. Others, for example, the fractions from crop residues that we put back into the soil each year, do change. We can manage crop residues, or alfalfa roots, or an alfalfa crop that is plowed into the soil. We can manage these sources of organic matter as they break down and form binding agents. That sort of binding agent may be pretty transitory and may last a couple of years and then disappear. But farmers are constantly managing the flow of such materials through a soil. You can provide a constant supply to the soil, so it will remain in a pretty viable condition.

Soil Structure and Compaction

BENBROOK: *What are the benefits of that soil staying in a more aggregated condition? Does it improve water-holding capacity, for example?*

LARSON: Yes, a soil's organic matter fraction has an influence on water-holding capacity, but the kind of changes that we can make practically in a soil are so minute that they do not greatly affect the amount that can be held, unless, of course, the thickness of the soil horizon itself is reduced over time by erosion. Management and soil structure certainly do affect, however, the amount of water that will infiltrate the soil and the rapidity at which water is transported through the soil. So, while soil structure affects runoff, erosion, and transmission down through the soil, it really does not affect water-holding capacity. That concern has been overemphasized.

Soil structure also affects the air, the gaseous composition of the soil. In essence, the soil atmosphere in a good soil is just about like the ambient air: 78 percent nitrogen, 22 percent oxygen, and .03 percent carbon dioxide. But in a soil that is compacted so that the air can't diffuse in and out of the soil, the carbon dioxide content can go up dramatically and the oxygen level down dramatically. If that happens, roots will be less healthy and you get into all sorts of problems.

BENBROOK: *A compacted soil is less porous, and you will have more surface runoff and slower infiltration?*

LARSON: Right.

BENBROOK: *So it is not that the soil cannot hold as much water necessarily, but that the water does not get into the soil; and, further, that compaction can restrict root growth. To the extent that soil aggregates are formed and bulk density goes down, compaction will be lessened and infiltration increased. Is that correct?*

LARSON: Yes, right. One of the real problems in modern agriculture, I think, is compaction. Soil compaction is becoming more and more of a problem. It is difficult to diagnose and difficult to measure exactly. While we can measure it in a certain field, we can't measure it in every soil in Minnesota. There is no simple, standard test. But I think most authorities would say compaction is becoming more and more of a problem, partly because of larger machinery. The harvesting equipment in many fields produces the greatest pressure on the soil surface.

BENBROOK: *And, of course, when harvesting equipment goes across a field in the fall with a little bit of moisture in the ground, isn't a moist soil much more prone to compaction than a drier soil?*

LARSON: Yes, a moist soil is much more prone to compaction than a dry soil.

BENBROOK: *Now, does compaction matter beneath the root zone?*

LARSON: Most of our work would indicate that the compaction from farm machines will extend maybe 20, at most 30, inches in the soil.

BENBROOK: *So if you have a compacted soil down to 20 or 30 inches and you moldboard plow, you are only going to affect the top 7 or 8 inches, and that can lead to what is called a plow layer.*

LARSON: Right.

BENBROOK: *The plow layer begins at the point in the soil horizon above which you have frequently tilled the soil.*

LARSON: It is the subsoil compaction that is the long-term worry. Compaction in the top 8 inches can be taken care of through routine tillage.

BENBROOK: *Corn roots and roots of a soybean plant will go down 36 inches, won't they?*

LARSON: Yes, probably more than that on a dry year. They will go down to 5 feet on a good soil in a dry year.

BENBROOK: *So that entire 5-foot horizon is there for the farmer to work with in terms of holding moisture and nutrients and feeding that growing crop. In the case of alfalfa, where the roots go down even deeper, it is farther down. There are subsoilers and heavy tillage machines that are available to break up a hard pan down 18 inches or 24 inches. Do these work well in breaking up compaction?*

LARSON: Subsoilers have a great appeal to farmers, but in most parts of the country they are usually a disappointment. They need to be used when the soil is dry, when they will shatter the soil. But if the soil is wet down there and you run that subsoiler through it, you may do more harm than good. You may actually shear the soil and compact it further. You may dig a trench through it, but you compact the rest of the soil.

BENBROOK: *Given the experience and effectiveness of using a subsoiler to deal with compaction, does the farmer have any other physical or mechanical means to get down into the soil to deal with the problem of compaction?*

LARSON: Well, you almost have to go to some deep-rooted legumes. Alfalfa we know can have a big influence under the right conditions. In the South, Kudzu and other weeds and crops can have a beneficial influence.

BENBROOK: *What can an alfalfa field over 3 or 4 years do to combat compaction?*

LARSON: The alfalfa plant roots just dig channels through the soil. The tap root of an alfalfa plant can be a millimeter or so in size. It has the ability to force its way through a compacted soil; when the crop dies, the root dies. That channel will still be there. Over time, these channels and aggregates that form in and around the channel will loosen up that subsoil.

BENBROOK: *What if you are putting a lot of organic matter back into the soil, working it in, so there is a lot of biological activity in the soil? Does biological activity affect the deeper parts of the soil horizon and reduce compaction?*

LARSON: No, I don't think that occurs too much.

BENBROOK: *So it is largely the action of the roots. What about worms? Do they play a role?*

LARSON: Certainly, they do. There is a lot of interest now in earthworms. Several of our people here at Minnesota are authorities.

Worms are animals, they have to have moisture. They need a source of energy. If we leave crop residues on the soil surface, the residues provide some of the energy, keep the soil cooler, and moderate extremes of temperature and water. People have measured the channels left by worms through the soil. The action of worms can be enhanced by a large amount of organic material near the surface or in the plow layer because they provide the food that the worms need. And these channels can have a tremendous influence on water transport and, like the alfalfa root, reduce compaction.

BENBROOK: *How deep can worms go? Two or 3 feet?*

LARSON: Yes.

BENBROOK: *So they can get almost down to the end of your normal root zone.*

LARSON: Depends on the soil; they can go deeper than that in some porous soils.

Biological Basis of Soil Productivity

BENBROOK: *Over the years there has been much discussion and research on what is going on in soils that sustain fairly high levels of corn or soybean production on farms that have not used heavy applications of synthetic fertilizers. Do you feel that we have a good understanding now of what biological processes are underway in the soil that can make, under certain management systems, relatively more of the 10,000 pounds of nitrogen in the top foot of soil available to plants? Do we understand why or how that happens on certain farms?*

LARSON: We don't understand it as well as we could. Soil biology has been a bit neglected in the last three or four decades because of our emphasis on commercial fertilizers. Nitrogen can come from many sources. It can come from manure, legumes, or free-fixing organisms. There is even a certain amount, 5 to 10 pounds a year perhaps, brought down by rainfall. In some cases, farmers use sewage sludges as a source of nitrogen. We probably understand the basic mechanisms we can use to work with each of these sources of nitrogen, but we have not really put them all in proper perspective.

The other thing to remember is phosphorous, too, is found in the soil in organic combination. There is mineral phosphorous in the soil, but there is also organic phosphorous in the soil. In native prairie soils in Iowa, about half the total phosphorous is actually an organic combination. And that has to be mineralized by the microorganisms also to make it available to a crop. We really understand that process very little. There has been pioneering work by Dr. Charles Black at Iowa State, and others, but still that process is not understood very well.

BENBROOK: *What does the energy available within soil for microorganisms have to do with soil fertility or corn production? What is the benefit of adding molasses or sugar to some soils?*

LARSON: If you add molasses or sugar, or any easily oxidizable organic material, you will get a tremendous growth of microorganisms. You can count the numbers in laboratory experiments; you can measure the oxygen that they consume or the carbon dioxide that they give off. You get a tremendous flush of growth within the soil. In doing so, it probably also influences the rate of oxidation of the native organic matter. This is often called the priming effect. So if you add sugar to a jar of soil in the laboratory, you will not only get that sugar decomposed quickly but you will also get a little more of the native organic matter. More than that I really don't know.

BENBROOK: *Does the increase in the rate of organic matter decomposition affect how much nitrogen becomes available to the growing plant the next season?*

LARSON: Yes, it will over the short run. You know how much nitrogen there is in the soil in all forms. It will speed the release of nitrogen, but you cannot make nitrogen out of nothing. It can be an important effect.

Coming back to phosphorus. Phosphorus recommendations for fertilizer applications by farmers are all based on inorganic tests. The so-called Bray test or the Olson bicarbonate test are based on the amount of inorganic phosphorus. Neither takes into account that organic fraction that might be mineralized. So, in systems where there is a lot of organic matter and microorganism activity, our standard soil tests are probably not very good.

BENBROOK: *Probably leading to over-recommendations on phosphorus?*

LARSON: It could. I think people are realizing this, but I don't think anybody understands it.

BENBROOK: *Let's go back to our two hypothetical farming systems. The rotational system with no intensive chemical applications and the conventional system where we are putting on a 150 pounds of anhydrous ammonia nitrogen in the fall and we are using herbicides for weed control and a soil insecticide for rootworm control. What impact does the injection of anhydrous ammonia at fairly standard rates, between 100 and 150 units per acre, have on the microbiological life in the soil?*

LARSON: Well, it doesn't have as much as some people would lead you to believe. Obviously, there will be a certain stimulus. Anhydrous ammonia has to be converted to nitrate, or will be converted to nitrate before much of it is taken up by the plants. That is a microbial reaction, so there would be a certain stimulation there. There is always some concern that applying anhydrous ammonia actually kills off the microorganisms because of the high pH that results, and this is probably true in a very small band—right within a centimeter or less adjacent to where that anhydrous is injected. But this is only very temporary, so over the long term it does not have much affect.

BENBROOK: *It doesn't affect earthworm populations?*

LARSON: It could. It could be toxic to earthworm populations. The big thing, however, in encouraging earthworms is adding lots of organic material back to the surface and keeping the surface of soils a little wetter and a little cooler, as you do under our residue management system. Worms have got to have an energy source, they have got to have some food, and their food is decomposing organic matter.

BENBROOK: *Well, surely with a continuous corn system or even a corn-soybean rotational system, unless the farmer is taking the stalks off in silage, there is going to be a pretty substantial flow of organic material back into the soil.*

LARSON: Yes, that is true. But say you harvest a 150-bushel corn crop and have, say, 3 or 4 tons of cornstalk that you plow under. They can decompose pretty fast, and by the end of the season, they are pretty well gone. They will have very little effect the next growing season in keeping the surface soils cool and moist. So you still have got that problem of high temperatures and low moisture in the surface soil where remaining residues may be found. Those earthworm holes, to be fully effective in helping water infiltration, have to come clear to the surface.

BENBROOK: *What do we know about the impact of agrichemicals on soil microbiological life, including earthworms? Is it something that has really been studied very intensively over the last couple of decades?*

LARSON: My impression is no. There is a lot of increased interest lately in this area. It is surprising how these things go. The Europeans and the Australians particularly have been interested in earthworms for years. We in America kind of got hung up on chemical fertilizers and agricultural chemicals and paid no attention. Now, interest is coming back, as it should.

BENBROOK: *Do you think that most corn residue, under common management conditions in the Midwest, decomposes fairly rapidly and is nearly gone by the next year?*

LARSON: Yes. Corn residues are low in nitrogen. Corn residue will average about 1 percent nitrogen on a dry matter basis, which equates to about about 20 pounds per ton. When they are plowed under in a soil, unless microorganisms in the soil have an ample source of nitrogen, the breakdown will be slow. The organisms have to have nitrogen to build up the protein in their bodies. The organisms will take first choice over the crop in getting at available nitrogen, so the rule of thumb to prevent this process, which we call immobilization, is to be sure that there is an ample amount of nitrogen there for the crop and the decomposition process.

BENBROOK: *Are mircoorganisms the farmer's enemy if they just take up nitrogen?*

LARSON: No, I think microorganisms are not the farmer's enemy, but they have to be managed like everything else. We have to consider what they do. Thirty or 40 years ago, much of the cornstalks, much of the wheat stubble in Minnesota, was burned just because of this problem. Farmers knew that if they plowed under several tons of wheat straw they would have a lousy crop the next year. They did not have the fertilizer to compensate for it. So they burned it, which is a very ineffective way to handle organic matter. Microorganisms can be the farmer's ally and a big asset, but they have to be managed. They are kind of an unseen ally, and most of us don't really understand or consider all of these things in our management decisions and recommendations.

BENBROOK: *When was the last time, do you suppose, that the Minnesota or Iowa Cooperative Extension Service put out a bulletin on managing microorganisms?*

LARSON: It has been a long time. It is a good thought, though.

BENBROOK: *Isn't it probably true, Bill, that to develop the science base to put out such a publication we are really talking about several years of new research to understand how to turn those microorganisms into an ally? We have evidence suggesting that some farmers have found out ways to do it, but we really don't understand, still, what is going on. We can see that it is working, but we do not know why.*

LARSON: I think there is no question that soil microbiology goes through cycles. Back in the 1920s and early 1930s we paid more attention to soil microorganisms, soil microbiology. Then came fertilizers after World War II. Fertilizers became so common and so cheap we had a real revolution in farming methods; we forgot all about these other things. Now we are coming back. Hopefully, some day, we will reach a proper balance.

BENBROOK: *Over the years, when synthetic fertilizers and pesticides became the dominant way to provide nutrients to plants, control weeds, and protect plants from insects, there has been a small group of innovative farmers and scientists and conservationists who have spoken of feeding the soil rather than feeding the plant. It is a lot more complicated, isn't it, to understand how to feed the soil rather than feed the crop?*

LARSON: Yes, it probably is because you can see the crop, you can sample it, you can study it, it is something visible. Microorganisms—there are millions of them in the soil, a lot of different species. They require a lot of different environments to grow and thrive. One may grow for a while, then use up its energy source and another one will come along. It is extremely difficult to know what is in the soil because the only way you can really identify them is to get them out of the soil. As soon as you extract them from the soil they change. So you are never sure that what you extract from soil is what was there and growing in the soil. I think perhaps it is fair to say, too, that microbiological research has generally been done in a laboratory, in part because it is so complicated. We need to get out in the field more and study some of these phenomena under field situations. And that is probably going to take a lot of development and new analytical methodology, as well as just plain hard work.

BENBROOK: *Is this the study of rhizosphere dynamics? Is that the right word?*

LARSON: Yes. This is the study of rhizosphere dynamics. Again, this is an extremely complicated system because you are talking about the chemistry and the microbiology and the interactions with the plant right at that root surface. You are talking about just a few millimeters or maybe less than a millimeter distance from the root.

The Establishment of Research Priorities

BENBROOK: *Back in 1981 you led an effort by 10 scientific societies to assess soil and water resource research priorities. The top priority that came out of that effort was sustaining soil productivity. Was the 1981 notion of sustaining soil productivity comparable to the dominant ideas about sustainable agriculture today in the soil science and agronomy research community?*

LARSON: It has changed a great deal. We were talking about sustaining a soil's productivity at that time. We were largely concerned with the loss of soil productivity from soil erosion. Now we are talking not only about erosion but a much more complex system. They tie together. One has evolved into the other.

BENBROOK: *Is it easy for a scientist to go from a concern over the impacts of erosion on soil structure and soil fertility to the impacts of farm management on a broader array of objectives?*

LARSON: Many of the same concepts apply. We, as scientists, are commonly criticized for being too narrow, for looking at our little area of expertise, and forgetting to look at all the rest. I think that is a valid criticism at times. I think the sustainable agriculture advocates and some of those involved with the LISA [Low-Input Sustainable Agriculture] program are saying just that. We have got to look at the whole picture, not just one little facet.

Role of Multidisciplinary Research

BENBROOK: *I understand that you have even been known to write a paper with an agricultural economist from time to time?*

LARSON: Yes, I have worked with agricultural economists a great deal. I have enjoyed interdisciplinary research. You learn from each other. Any researcher who isn't a student, doesn't want to learn something new, is no good as a researcher. Any educator and researcher should constantly be on the lookout to learn something new, and interdisciplinary research is one of the ways you learn new things.

BENBROOK: *Bill, has the majority of the faculty here in the soils department at Minnesota, or the majority of your colleagues still working in ARS, participated in and benefited from multidisciplinary research activities to the extent you have?*

LARSON: I think in recent years there has been a lot more emphasis on interdisciplinary research, at least here at Minnesota. Partly because of Dr. Gene Allen, our acting vice president for agriculture, forestry, and home economics, and others, we have come a long way in encouraging this and getting the job done.

BENBROOK: *How have research leaders here at Minnesota tried to provide more opportunity and encouragement for multidisciplinary research?*

LARSON: At Minnesota, we established a Center for the Agricultural Impacts on Water Quality. This covers about five or six different departments. It includes soil scientists, crop scientists, agricultural engineers, plant pathologists, entomologists, and economists. In this particular case, because we created this center and got some recognition for it, the state legislature provided some new monies. Dr. Jim Anderson from our department now leads the center. It is really the best example of an interdisciplinary team I can think of that came together, not because they were ordered to, but because they wanted to. There was a problem that needed a new approach and some administrative leadership. I think that is what we have got to do more and more of.

BENBROOK: *You spent several years on campus here developing the erosion-productivity model and a new scheme for characterizing midwestern soils. Your work has played a major role in shaping, and now is the basis for administering, Minnesota's RIM program [Re-invest In Minnesota, a state-level program to expand enrollment in a CRP-like program]. During that time, in developing these models and information, you worked with several colleagues on campus, including Ford Runge, an agricultural economist. Did you and Ford ever receive Hatch Funds to support any of this work?*

LARSON: No, not really. The budget system at most universities is that each professor or each investor/researcher gets a certain amount of money and, within limits, is kind of free to spend that the way they want to. Dr. Runge and I worked together on foundation grant monies because we wanted to work together. We recognized the challenge, we had some common interests, we realized that by pooling our talents we could get a lot further than if we tried to do it alone.

BENBROOK: *So the work that you did with Ford really did not require an outside source of funding to facilitate it.*

LARSON: Well, it did not require it, but outside funding would have moved it along, and we eventually did get a grant from the Northwest Area Foundation and some other grants. A lot of that work, however, was bootlegged.

BENBROOK: *In bringing together scientists to do multidisciplinary research on major problems, like water quality, do you think a university has an easier time of overcoming the hurdles imposed by disciplinary boundaries and turf consciousness, or does the ARS have an easier time? Which component of the research system—academe or ARS—has the capacity and ability to recombine its available resources more fluidly to address problems or respond to scientific opportunities?*

LARSON: I think there are two ways to look at that. The university has an advantage because we have in our College of Agriculture 10 different departments, all here on the campus. We all know each other, we all see one another. We have much easier access to the human talent needed to do a multidisciplinary study in academia than is often the case in ARS. ARS scientists tend to be located by disciplines, and all of the disciplines needed to address a problem may not be at a given location.

In terms of funding, perhaps ARS has the advantage. I think the agency is a little more willing to move its resources around to where the most pressing problems are.

BENBROOK: *When you are working in the Soil Science Department and a colleague from another department walks in the door and proposes to you an interesting area for collaboration, you don't feel constrained to have to get approval from the dean or consider the budget implications—you pretty much just do it. Is that how it happens?*

LARSON: Yes. Any responsible professor or faculty member will try to keep his department head or his dean informed, but usually he is not constrained. If he has a good idea, a good project, and wants to work with someone else, he is usually encouraged.

BENBROOK: *Now, scientists in ARS go through a somewhat more formal and prescribed annual plan of work. Does an ARS scientist feel somewhat more bound by that plan of work and less free to pursue new research directions?*

LARSON: Yes, I think he is more constrained by what is written on paper in his plan of work.

BENBROOK: *But within ARS from year to year, the content of a scientist's plan of work might change more fluidly than the job of a professor of soil science at Minnesota. A professor usually has joint responsibilities between teaching and research.*

LARSON: Yes, that is the way it works.

Shifting ARS Research Priorities

BENBROOK: *If you were managing the USDA's agricultural research effort, what areas of research would you expand, even at the expense of existing programs?*

LARSON: If I were administrator of ARS, I think I would take the attitude that ARS is a national organization dedicated to meeting national needs. It has the opportunity and the responsibility to provide national leadership, and because it is a big, national organization, it is structured to move from one problem area to another a little easier, on a national basis, than academia.

BENBROOK: *When we started out, you noted four pressing natural resource management challenges: better use of land and too much land in the wrong use; environmental problems, such as water quality; sustainable agriculture; and the economics of agriculture. So, if you had the opportunity to participate in shifting priorities within ARS, research addressing these problems would be among the areas receiving increased support. What are some of the areas of ongoing research that you feel can be cut back now to free up some resources for redirection?*

LARSON: I think probably our soil fertility research can be redirected. I am not saying that it should be dropped, but I think we need to look at it much more broadly than just chemical fertilizers. Some resources should be moved into low-input agriculture, sustainable agriculture, or whatever we want to call it.

We need more soil biology research, particularly field-type studies where you are looking at the dynamics of the system in the field.

BENBROOK: *You don't think the biotechnologists, then, are going to answer the critical questions of soil biology and what is going on in the soil? You think it is going to take some field-level work, too?*

LARSON: There is a whole field now that gets a lot of attention called microbial ecology. The best geneticists and biotechnologists in the world cannot breed a super new microorganism unless they know something about what kind of an environment that microorganism will grow in. We have a lot of superior strains of, say, rhizobia, that fix nitrogen with legumes. The problem is that many species won't survive in soils. They are taken over by the native strains because we do not quite understand what the effects of the environment are. So while biotechnology offers lots of promise, it is not going to answer all of our problems.

BENBROOK: *The work that you, Dr. Gupta, and others did on energy production from biomass and the potential consequences for erosion control of using midwestern and Great Plains crop residues for energy production, how did the decision get made for that work to be done and where did the resources come from? How was the leadership provided to coordinate that multidisciplinary effort?*

LARSON: That is an interesting case. I got interested in it because a colleague here at the university was serving on a state task force in which energy supply was a major concern. That was when Middle Eastern nations were threatening to limit oil exports. My colleague kept asking how can we determine which soils could tolerate removal of crop residues and which ones could not? Answering this important question was a new challenge. I hit upon the idea of using the national data base that SCS had provided, combining it with data on nutrient content of residues and other factors to come up with an assessment. We did it for the whole Corn Belt.

I recall very vividly, at a reception one night, I cornered our administrator, T. W. Edminister, and told him about this work. He said, "You come into Washington in a couple of weeks. I want to talk to you more about this." As a result, he did give us some extra money. That was how I hired Dr. Gupta initially to do the work; he was a computer expert. We thought it was so successful in the Corn Belt, we decided to look at the rest of the country, make this really a national assessment. And so Mr. Edminister told me to go ahead and organize the activity. I contacted colleagues in the Southeast, Northwest, and other parts of the country.

BENBROOK: *Did the collaborators at other campuses require some research support to do the work? Or did they just bootleg it out of their ongoing sources of support?*

LARSON: I think they got a little help, but most of it was probably bootlegged. We had some limited support, but I can't remember how much. Mr. Edminister recognized it as a good approach and helped make it possible.

Environmental-Economic Linkages

BENBROOK: *What about the linkages between a farming system's economic and environmental performance? Do you feel that environmental and economic linkages are important and real?*

LARSON: Oh yes, certainly, they are important. Sometimes they seem in conflict. Erosion is a good example. We have had reports that leads one to believe that sound conservation farming is not economic. I think that conclusion results from looking at only part of the problem. Again, it is a problem of scientists working in isolation and not understanding the full range of impacts that arise from erosion or choices in management systems.

BENBROOK: *Over the years there have been many efforts to quantify the impacts of erosion on soil productivity. Efforts have been made to estimate the economic value of a ton of soil, efforts to determine the intergenerational interests in soil erosion, to build an economic dimension into T values, etc. You and your colleagues have made important contributions to this work. In general, though, isn't it fair to say that these models, including your own, are generally unable to capture or measure empirically some of the more subtle but important linkages between resource quality, stewardship, and economic performances?*

LARSON: Yes, I think that has been one of our big disappointments. We haven't been able to do it. I think, probably, we just have not put enough effort into it, although it is an extremely complex problem.

BENBROOK: *Some of the agricultural economists that have worked on these questions have reached the conclusion that there is really nothing there. That because they haven't yet been able to measure important linkages, they must not exist. Are you still skeptical about that?*

LARSON: Yes, I am skeptical. I am not sure why we have not been able to develop models that more fully capture what goes on in the field. I think part of the problem is that we have not worked together enough. I recall a number of years ago that a very noted national economist told me that he thought that this whole matter of soil erosion was a bunch of bunk because on his home farm in Illinois crop yields had gone up every year for the last 50 years despite continuing erosion losses, so why were we concerned? It is much more subtle than that and much more complex. Our current understanding fails to take into account many hidden, often long-run costs. Natural resources will be around here for a long, long time, and we will need to use them more and more intensively, not just for food and fiber production, but also to enhance the environment of Minnesota—the camping, recreational facilities, woods and forests of Minnesota, the wildlife. All these things are important. We have got to look at them all together and be mindful of how resource management today may influence the options we pass along to future generations, who are likely to have even more trouble than our generation in balancing production and environmental goals.

The Severity of Erosion

BENBROOK: *In your 1983* Science *paper, "The Threat of Soil Erosion to Long-Term Crop Production," you and your coauthors, Drs. Pierce and Dowdy, wrote: "Thus the debate as to whether soil erosion on cropland has increased or decreased since the 1930s cannot be resolved." What can be said in 1989 about the extent and seriousness of soil erosion today in contrast with 10, 20, or 50 years ago? Do you still feel that you cannot say much with any degree of certainty?*

LARSON: I think erosion remains a very serious problem on perhaps some 20 percent of our cultivated land. It is clearly not a serious problem on all our land, although it may be somewhat of a problem on most. As I indicated earlier, with conservation tillage practices and some residue management practices we can control erosion pretty well.

BENBROOK: *But given that the physical properties of soil in the Midwest were the same back in the 1930s as they are today, and correcting for how crop management systems have changed and the increase of corn yields from 40 bushels to 120 bushels on good soils, could one retrospectively estimate past erosion rates based on the agronomic, tillage, and cultural practices common in the 1930s?*

LARSON: I think it could be done. In the 1930s, I mostly remember western Iowa and Nebraska. We planted our corn to get 8,000 or 10,000 plants per acre. Now we criticize our farmers if they don't get 25,000 plants per acre. Those 8,000 to 10,000 seeds planted in 40- or 42-inch rows did not give the soil good canopy protection from beating rains. Those 8,000 stalks with a 40-bushel yield probably produced about a ton or a ton and a half of residue. Today, if we don't get three tons we are disappointed. So we did not have the residue to put back in the soil or maintain on the surface.

BENBROOK: *What about weed control?*

LARSON: Weed control was a real problem. In Nebraska, where I grew up, we always cultivated corn three times. We thought that "was written in the Bible." Here in Minnesota, in the quack grass areas and so forth, they cultivated four and five times. Every time you cultivate soil you lay that soil open and expose it to erosion.

BENBROOK: *Bill, do we need a more accurate universal soil loss equation? If so, why? And how?*

LARSON: We always need more accurate estimates. However, I think perhaps we [in soil erosion research] have been a little hung up on predicting erosion and not concerned enough about the effects of erosion on crop productivity, its off-site damages and so forth. We get overwhelmed with the prediction process and don't spend enough effort working to assure erosion rates are reduced through sound conservation.

BENBROOK: *What about T values? Are current T values defensible, given the new importance of T values in the erodibility-index and in the administration of the conservation provisions in the 1985 farm bill?*

LARSON: I think it is really difficult to try to express all the effects of erosion in one number called a T value. There are environmental concerns, ecological concerns, and crop productivity concerns. One simple number cannot express them all. I don't quarrel with what they were used for in the past, but for many of the concerns today they are inadequate.

BENBROOK: *In the 1950s and 1960s, when the memory of the Dust Bowl days were still strong in peoples' minds, the dominant concern was how much erosion can a soil tolerate without having its long-run production impaired and how to reduce erosion rates. In the decades ahead, do we need some new concepts and benchmarks in the area of water quality protection that can play a role similar to the T value in erosion control?*

LARSON: Yes, I am sure we do. I think we have got to learn how to farm by soils rather than fields. We have to learn to manage variability in the landscape. I see a need for some concept like a T value for water quality, but I'm not sure of how to express it.

BENBROOK: *Wouldn't one option be some function of the nitrogen flow through the cropping system and the efficiency of nitrogen uptake by the plant. The more nitrogen is taken up by the plant relative to the nitrogen that is available, the less nitrogen there is to leach into the groundwater. This could evolve into a measure of the portion of nitrogen in the soil that is potentially available and likely to leach down into the water.*

LARSON: Yes, certainly. There is a delicate balance between too little and too much nitrogen. No life can grow without nitrogen. I think we perhaps have been a little bit overwhelmed with the thought that we want to get the last bushel of corn we can off each field. To do so, we put on a little more fertilizer, just to be sure the crop gets enough.

BENBROOK: *Because nitrogen has been so inexpensive and because we really did not understand any substantial adverse consequences for the environment from being sure there was plenty there, farmers have, in the 1970s and 1980s, assured that nitrogen is rarely the limiting factor in crop production. As a result, farmers have been achieving yields pretty consistently near the maximum that probably is feasible, given moisture availability and genetic potential. Do you think that the current concerns with high-input costs and environmental impacts of farming systems could change things profoundly enough so that, perhaps 30 years from now, farmers would manage their crops and soil fertility so that they are shooting for the highest possible efficiency in production, given average weather expectations and given available soil moisture?*

LARSON: We can fine-tune our nitrogen management much more than is currently the case. We are not going to stop using nitrogen, but we may use less. It is mostly a matter of fine-tuning. Much of that research to do the fine-tuning has been done. Now, it is a matter of extension education and making better use of modern soil surveys, computers, etc. We may have to think of a whole host of alternative systems. One system for the small farmer, another one for the large one.

BENBROOK: *Scientists are always seeking better data, attempting to refine models, and develop more complex and robust empirical characterizations of natural events and processes that are random and complex. Policymakers, on the other hand, are inclined to seek simple, practical answers regarding the nature, geographical distribution, causes and consequences of agricultural management, and land use practices. They want something simple—what's the problem, where is it, and what do we need to do to fix it? You have worked both as a research scientist and in the policy arena trying to answer simple, practical questions, yet you also understand the complexity of the underlying issues and the kinds of tools farmers will need to translate scientific knowledge into practical solutions to common stewardship challenges. Where do we need to focus in the future? Do we need better models and more data, or do we need more practical analyses and better application of existing data and models? Sounds to me like you are leaning a little bit toward better use of existing information.*

LARSON: Yes, I hesitate to say we need one more than the other, but I think the opportunities for using existing information are really great. We are not using all we know. Some of the analyses we did for the RCA assessment and some of our work on energy and crop residue removal were attempts to demonstrate that a mass of data could be combined and used creatively to address new questions of national significance. I think there is a lot more that we could do. In defense of the university scientists, this sort of analysis is not encouraged. I hesitate to say this but our whole tenure and promotion system and the promotion system in ARS is based on publications. Publications in scientific journals tend to focus on discrete, well-defined research on some small aspect of a complex system, not the generally less conclusive results of multidisciplinary research efforts. I think that is unfortunate.

BENBROOK: *Bill, thanks very much for sharing so candidly your views on these issues. I know you will continue applying your great skills with people and your experience and scientific understanding to key resource stewardship challenges, and for that the nation is grateful.*

13

Soil Management Research in the Search for Sustainable Agriculture

C. F. Bentley

I have been requested to present concluding remarks for this book, providing a synthesis of it, and to give futuristic views on research and development priorities as I see them.

The space allotted for these remarks makes it impractical to attempt a review of the individual chapters. This chapter touches on a highlight of each, however, and presents an expanded version of the suggestions I am going to make regarding research and development priorities.

Soil Management is Fundamentally Important

This book is concerned with soil management and sustainable agriculture. The chapters discuss many aspects of soil quality, or effects on it, as affected by soil management, which has the potential both to degrade and to enhance soil productivity. Thus, appropriate soil management is a fundamental requirement for sustainable agriculture.

These chapters were arranged to represent, at least in part, the great diversity of Dr. William Larson's interests and research activities. They illustrate how his wide interests and great productivity as a researcher have contributed to improved understanding of soils and the importance of preserving their productivity under agricultural use.

Yet, in spite of Dr. Larson's prodigious research outputs, the chapters make it clear that much remains to be learned about soil management for sustainable agriculture. Research funding for soil management and on soil quality is inadequate in relation to their fundamental importance to sustainable agriculture.

I think the global publicity regarding the remarkable contributions of the high-yielding crop varieties has been a hindrance to the other agricultural sciences, especially soil science, because financial priority seems to have gone to plant breeding and related activities. Most nonagrologist policymakers and financial managers appear to be unaware of the fact that, however remarkable the potentials of high-yielding crop varieties may be, attainment of their production potentials is very dependent on soil quality and soil management. Applied research on soil management and the maintenance of the productivity of agricultural land under use has been, and is, seriously underfunded.

But I would add that the nonanswers to some of the questions raised in this book, questions related to the nonadoption and economic realities of some research outputs, constitute challenges to us. The general citizenry and farmers are increasingly concerned about the environment and the economic aspects of sustainable agriculture. If there is indeed inadequate support for research on soil management and sustainable agriculture, is it, in part, because we have not been doing sufficient on-farm validation of research outputs and determination of their acceptability to farmers, who must be economic realists to survive?

Highlights from the Chapters

My selections of some highlights from the chapters follow:

Soil Erosion and Crop Productivity. Elliott and associates and Pierce briefly review some history of efforts to assess the effects of erosion on crop yields and soil quality. The goal was to develop the capability to predict effects of erosion on production. Larson's work in this area contributed to the basic knowledge needed for planning soil conservation programs and policies. But much research is still needed on the complex soil management problems entailed in practical reductions of soil erosion. The authors provide a list of challenges still to be met. Their resolutions will ensure that Larson's students and others will have much to do in the years ahead.

However, the authors conclude on a positive, optimistic note: The pioneering work of Larson and others justifies optimism that the soils of the United States will be able to meet the agricultural production needs of the next century.

Soil Structure. Edwards and Groenevelt state that farming attempts to manage soil structure in order to enhance its effects on crop production and on the movement of water and chemical in soils. Structure is affected by machine traffic and tillage implements, the growth and decay of plants, crop and soil management practices, and climatic characteristics.

Research by Larson and Gupta has contributed to an understanding of some of the complexities involved.

Edwards and Groenevelt state that soil structure and porosity in nontilled topsoils is very different from that of annually tilled soils. Earthworms and the persistence of their burrows in nontilled soils may affect infiltration and movement of chemicals in soils with a dense soil structure.

Soil Compaction. Cruse and Gupta discuss soil compaction. They state that, in general, compaction reduces saturated water flow, slowing water drainage, restricting root growth, and reducing diffusion of gasses. Lower yields usually result.

Most compaction is said to be an unintended effect of farming operations. Increased equipment size and power tend to increase compaction problems because operations are possible on soils that are too wet, and greater axle loads cause compaction at deeper depths.

Dr. Larson's compaction-related research has been primarily concerned with prediction of responses to load or stress application. He sought understanding of soil bulk density under natural conditions as a prelude to determination of how soils respond to applied forces. The practical results led to recommendations that field loads not exceed the critical stress load.

Cruse and Gupta conclude that Larson has contributed scientific understanding and modeling methodology for describing and solving the compaction phenomenon. His research has received international recognition both from applied as well as more basic scientists in the field of tillage and compaction. He has helped to identify basic soil characteristics that can be successfully applied to predict field applications.

Adoption of Conservation Tillage and Associated Planting Systems. Allmaras, Langdale, Unger, Dowdy, and Van Doren state that the current American working definition of conservation tillage is "a tillage-with-planting combination that leaves at least 30 percent of the soil surface covered with residues after planting." In the United States, the percentage of planted cropland so managed rose from 25 percent to 32 percent between 1980 and 1988.

The authors relate the results of a 1985 survey in the 11 tillage management units of the United States that appraised benefits, problems, and other considerations related to conservation tillage. In spite of understandable variations among practices in the various tillage management regions, they conclude that all showed improved technology supported by increased production or lower production costs.

The authors refer to Larson's contributions to these aspects of soil management.

Management of Agricultural Land for Receiving Wastewater Sludges. Dowdy, Page, and Chang discuss land farming of sewage sludges and other effluents as a means for their safe disposal and as sources of nutrients. Possibilities of chemical contamination buildup and of nutrient imbalances are concerns.

Data from a case study with municipal sludge near St. Paul, Minnesota, are discussed. Over a 10-year period, yields of corn and reed canarygrass on sludge-treated and fertilized areas were compared. Dry matter production was excellent and comparable on the two areas. Monitoring to depths of up to 1.5 meters found no cause for concern because of changes or accumulations of heavy metals or nitrates. Larson's involvement in this project included his active physical participation in the field work.

Natural Resource Assessment. Benbrook's chapter is unusual in that much of the material for it was obtained during a taped, day-long interview with Dr. Larson, who had done extensive work on policy-related issues. Analytical procedures developed by Dr. Larson and some of his associates had enabled extraction in useful forms of much information from the Natural Resources Inventories of 1977 and 1982. That knowledge contributed importantly to some of the policies in the 1985 U.S. Food Security Act. Similarly, Dr. Larson's work had a role in the recognition that soil productivity should be concerned with the effects of soil management on the environment and sustainability, in addition to being concerned with current yields. He also brought attention to the problem that some land is being used for types of production that are not compatible with soil characteristics.

Sustaining the Resource Base of an Expanding World Agriculture. The chapter by Stewart, Lal, and El-Swaify is primarily concerned with how agricultural practices and systems affect the sustainability of the land resource base in a global context. They explain that a truly sustainable cropping system is one in which, over the long term, the effects of soil-conserving and maintenance practices equal or exceed the effects of degradative processes.

After an extensive discussion of soil, climatic, and social conditions, which are affecting soil productivity and the global environment, the authors ask, "As soil scientists, do we have a major role to play?" They then propose several roles, including:

► Realistic assessments of the potentials of various soil resources and their environments.

► Performance of applied site-specific research to assess effects of recommended production practices (technologies) on the land resource and, thereby, a role in promoting technology transfer by such validations.

► Involvement, as appropriate, with other scientists in efforts to overcome constraints that inhibit or prevent adoption of production practices that would contribute to sustainability.

► Compilation by research of more complete characterizations of the soil resources with suitability for agriculture.

Stewart and associates emphasize that arid and semiarid regions are the ones where soil degradation and environmental damage are of greatest concern. I appreciate greatly these authors' emphasis on the fact that livestock-related matters are inadequately considered and supported in relation to agricultural sustainability and environmental problems in developing countries.

Energy Flow and Sustainability of Agricultural Systems. Pimentel and Heichel introduce their challenging chapter with a statement that I think all agrologists should stress at every opportunity in order to educate legislators and consumers alike.

"An adequate **and sustainable** [emphasis added] supply of food is basic to the survival and progress of human civilization."

The authors then state that contemporary, highly productive, energy-intensive agricultural systems are of questionable sustainability for three reasons:

► They depend on fossil energy, which is not sustainable.

► Serious soil erosion accompanying contemporary agriculture limits sustainability of such systems.

► The economic stability of contemporary agricultural systems is uncertain.

Pimentel and Heichel state that earthly resources can support a world population of about 1 billion at the contemporary American level of living. With careful planning, it will take about 200 years to reduce world population to that number. Therefore, they say there is an urgent need for technologies and policies to attain stable populations and environmentally sound, sustainable agricultural production systems.

Those provocative views are supported by data comparing energy output/input ratios for a variety of agricultural systems. As presented, the examples show that contemporary, energy-intensive agricultural systems have the least favorable energy output/input ratios and are characterized as being the most soil-degrading systems considered.

Ethanol as an alternative to fossil fuels is not practical, they conclude, because its manufacture requires more energy than is produced.

Legume-based crop rotations, preferably including forage crops, are depicted as the most attractive production systems on the basis of both energy and environmental considerations. The authors state that, unfortunately, contemporary U.S. commodity programs discourage farmers from adoption of legume-based crop rotations, which are economically viable and/or of benefit to society.

This chapter generated much discussion.

Soil Data Bases for Land Use Decisions. The chapter by Larson and Robert proposes that "prescription farming"—the basing of soil management operations in farm fields on the specific characteristics of the various soils in a given field—is now technically feasible and represents an important advance in soil management as well as environmental protection. For example, microprocessors are capable of changing the application rates of fertilizers and herbicides to conform to the optimum requirements for the various soil series and differing soil characteristics in a given field. However, to quote the authors, practical on-farm exploitation of the new technology will require "better soil maps digitized for rapid recall, for better landscape descriptions, and improved low-cost application equipment."

I would add that effective on-farm use will also depend on very sophisticated farmers.

Fundamental Neglects in the Search for Sustainable Agriculture

The symposium on which this book is based was organized by the World Association of Soil and Water Conservation in cooperation with the Soil Science Society of America and the International Affairs Committee of the Soil and Water Conservation Society. It is, therefore, appropriate and important to emphasize international aspects of sustainable agriculture, with a focus on soil management and environmental problems in many developing countries, which are of truly global importance.

Let me first discuss soil-conserving and productivity-maintaining needs in developing countries.

A report commissioned by the Consultative Group on International Agricultural Research (CGIAR) for presentation to that body at its May 1989 meeting in Canberra, Australia, identified the three "most pervasive threats to sustainable agriculture in relation to the overall priorities of the CG system" as:

► Soil degradation—"clearly the most significant threat to sustainable agriculture."

► Decline in genetic diversity—"the erosion or narrowing of the genetic base of agriculture."

► Pest and disease problems.

The report to the CG states that the above three areas account for more than 45 percent of all activities related to sustainability at the 13 international agricultural research centers.

In my opinion, there are two other gravely neglected areas for research and support that are of comparable or even greater importance to sustainable agriculture and soil conservation in developing countries. They are:

► Fodder legumes and grasses, for arable land cropping systems as well as for grazed lands.

► Woody species, for fuel materials, alley cropping, livestock fodder, wind shelters, farm use, household purposes, or whatever.

A Puzzling Imbalance. It is baffling that agricultural research and agricultural policies in, for, and by developing countries have so generally, outrageously, and consistently neglected the roles of the above two major factors with great potential for contributions to sustainable agriculture.

In a majority of developing countries, the livestock component of the agricultural sector accounts for 25 to 75 percent of the actual value of agricultural production. The general neglect of that sector is dramatically illustrated by the imbalance in projected needs for research scientists by the year 2000 for selected commodity and program areas as given in the World Bank's West Africa Agricultural Research Review (2). Therein, staff requirements projected for specifically listed crops (excluding forestry) is 36.3 percent of the total need for scientists, whereas the scientist's need for livestock production is projected as only 6.4 percent. Anticipated need for soil scientists amounts to only 5 percent of the total need. No need for forage crop or pasture scientists or for research on woody species (i.e., noncrop and nonforestry trees) was foreseen.

Adverse Effects on Soil Quality. Postel (*1*) has described the most serious aspects of land degradation on the global basis as follows:

"Each year, irreversible desertification claims an estimated 6 million hectares worldwide—a land area nearly twice the size of Belgium, lost beyond practical hope of land reclamation. An additional 20 million hectares annually become so impoverished that they are unprofitable to farm or graze."

I submit that lack of research on, and exploitation of the potentials of, legumes and grasses in crop rotations or for use as animal feeds, and of woody species for the purposes previously listed, constitutes one of the most serious oversights in the quest for sustainable use of agricultural land resources, whether for arable or pasturable agriculture, in developing countries.

I justify that statement because in many or most developing countries the accelerating removal of crop residues from cultivated fields for use as animal feed or household fuel, and the increasing use of animal dung for domestic cooking, together with overgrazing of unimproved, largely unmanaged grazing lands, are among the most important agricultural causes of soil degradation. Without residues retained in arable fields, organic matter content decreases, soil temperatures soar, crusting problems are aggravated, runoff and erosion accelerate, moisture infiltration is less uniform, biological activity (especially of earthworms) in the soil declines, soil compaction is favored, and soil degradation processes are progressively augmented. Additionally, failure to retain crop residues in cropped fields and the burning of dung for cooking adversely affect soil fertility because nutrients are not recycled.

Reality demands recognition that only an exceedingly small proportion of farmers in developing countries will have mechanical power for field operations within the foreseeable future. Most farm field operations and hauling will be done by animal traction or human muscle power. Although plowing to depths of 6 to 12 inches results in important yield increases on considerable areas, lack of sufficient feed for animals frequently makes such plowing impractical. Animal traction for seeding with drills and/or cultivation for weed control or other purposes are often hampered or prevented by inadequate feed for working animals.

Land shaping with contour ridges and furrows can be important in achieving improved moisture uniformity in fields and for preventing or reducing water erosion. However, in most developing countries, without animal traction, such desirable practices are seldom possible. Many legume fodders and numerous woody species have strong, deeply penetrating roots that can act as biological plows, opening up compacted soils for better infiltration of rainfall and penetration by roots of grain crops.

When used for intercropping with grain crops, some woody species and many legume or grass fodders can serve as living mulches in some agroclimatic zones. In such cases there is usually advantageous recycling (or pumping) of nutrients, as well as potential to exploit the phenomenon of biological nitrogen fixation.

Cooking-fuel shortages are escalating rapidly in much of the developing world. Currently, in India, income from production of fuel wood on previously cultivated land sometimes exceeds returns from growing food crops, and some such changes in land use are occurring.

Hampering Traditions. The failure to develop and use forage crops adequately for livestock feed, as well as the failure to exploit woody species (other than tree-crop and timber species) to their logical potentials in most developing countries, results from a variety of hampering social and cultural traditions.

Often, public land is the community grazing area as well as a source of wood. Normally, everyone has the right to use such areas, but there are no provisions for protection, maintenance, or management. Previously, with sparse populations, natural regeneration enabled sustainable production, at a rather low level. Currently, if there is no management or improved practices, population pressure usually results in degradation and reduced production. There may also be an inherent resistance to planned production, which entails work or inputs, for something that has traditionally been "free."

Time horizons for production of food crops are usually measured in weeks or months. Forage crops and woody species are usually slower growing, sometimes taking years to develop. Consequently, uncertainties about the future may deter investments of labor or capital that may be lost. Lack of proven benefits or returns under site-specific conditions is often an accompanying deterrent.

Livestock may be primarily scavengers, and the concept of producing feed specifically for them may be totally foreign and unattractive. Indeed, the reality is that the planting and production of crops to be used specifically and solely as feed for livestock is largely unknown in significant parts of some developing countries.

In any case, reliable economic data evaluating returns from investments of labor and resources for the production of woody materials for fuel, mulches, or other purposes, as well as the growing of legumes and grasses as feed for livestock, are exceedingly scarce or totally lacking in many developing countries. When research on forage crops or woody species for agricultural use has been done at research establishments or experimental stations, the conditions are usually so different from those that ordinary farmers must grapple with that research data and extension efforts based on experiment station research do not interest the farmers. Moreover, even when there are good, practical results from national agricultural research systems, the extension system may be incapable of successful promotion for adoption by farmers. Indeed, the world Bank has obtained generally discouraging results from its loans to fund extension projects in developing countries.

The Challenges. Worldwide, agriculture faces very tough challenges, many of which are land use related. The following are examples:

► Achieve increased, affordable sustainable agricultural production.

► Halt and reverse the trend toward degradation of agricultural land, cultivated and pastoral.

► Reduce environmental pollution resulting from excessive use, or through misuse, of agricultural chemicals.

► Cope with agricultural pests while increasing agricultural output.

► Control reasonably the adverse effects of runoff and erosion from agricultural land, as well as on the quality of ground, river, and other water bodies.

► Reduce destruction of forests in order to expand agricultural production.

► Protect the environment and the future through reasonable preservation of genetic diversity.

► Reduce human drudgery in agricultural production where it is practical to do so.

Temperate Regions Too! Soil quality deterioration and the sustainability of agricultural production under arable use in the more developed countries is now recognized as a serious problem.

Although residue-retaining types of tillage and zero tillage have shown merit in partly meeting the challenges listed above, such techniques are not universally suitable in temperate regions. Their economic practicality is also questioned in some areas. Moreover, in temperate regions, crop rotations, which include forage crops periodically in the cropping sequence, have declined markedly with the increase in contemporary monocropping and grain-only systems, such as continuous wheat or corn-soybean rotations.

I am of the opinion that there are some results of crop rotation research buried in long-forgotten files that would merit reconsideration and rigorous on-farm evaluation if we are to successfully meet the challenges of maintaining soil quality of cropped land.

The Bottom Line

I submit that a massive increase in practical, applied soil and crop management research aimed at greatly increasing the exploitation of woody species and a considerable increase in the use of animal traction and of planted production of forage crops has tremendous potential to solve, or at least mitigate, most of the challenges listed above.

I envision extensive programs of rigorous, on-farm validation research experiments executed jointly by competent researchers and extension personnel with the active participation of cooperating farmers. Such an approach will win farmer acceptance of appropriate recommendations and practices. Impractical recommendations will be rejected, and adaptive or new research will be needed to develop improved recommendations and methods acceptable to farmers. The Brundtland Commission, in its report *Our Common Future* (*3*), urges much more extensive use of such an approach to applied agricultural research.

Funding sources must be convinced that, regardless of the excellence of improved crop cultivars, attainment of their potentials is dependent on soil management practices and soil quality. Equally, it is pointless to continue to develop research outputs with potentials to increase agricultural production if those potentials are not going to be realized more adequately than they are currently because of soil degradation or because of a lack of effective programs for on-farm validations and to promote their adoption by farmers.

I think the leadership to exploit the potentials outlined is primarily a responsibility for soil and water conservationists in their role as practical integrators of multidisciplinary activities. The task is tremendously complex and difficult. The need and the potential dividends are enormous.

I conclude with two questions:

► Are we equal to the challenge?

► Do we have a sufficient number of Bill Larsons to exploit the potentials of what I am convinced are opportunities to make truly important contributions to the preservation or restoration of productivity on vast areas of the land resource on which we all depend?

REFERENCES CITED

1. Postel, Sandra. 1989. *Halting land degradation*. In *State of the World 1989*. W. W. Norton, New York, New York.
2. World Bank. 1987. *West Africa agricultural research review, World Bank, 1985-86*. Washington, D.C.
3. World Commission on Environment and Development. 1987. *Our common future*. Oxford University Press, New York, New York.

14

Soil Management in the 21st Century

F. J. Pierce and R. Lal

The world's arable land base is limited, and many hectares of once arable land have been degraded or converted to alternate uses in the 20th century. World population is expected to exceed 6.7 billion by the turn of the century, thereby continuing the rapid increase in pressure on soil resources to produce food and fiber. Food quality and safety, the farm economy, depletion of aquifers, nonpoint-source pollution of surface waters, leaching of contaminants to groundwater, and continued soil degradation are the fundamental problems facing agriculture in the United States today. Globally, two of the most important issues as we approach the 21st century are concern over the adequacy of the world soil resource base to meet food and fiber demands of an expanding world population and concern over the quality of our environment, including air and water as well as soil.

Meeting food and fiber demands in the next century will require higher productivity levels for land now in production, the addition of new land not currently in production, and the restoration of degraded lands to reasonable levels of productivity. Soil management practices in the 21st century must sustain and improve the soil resource base; contribute positively to the quality of air, water, and soil resources; and meet the economic and social needs of the food and fiber producer.

It is a difficult proposition to project what soil management systems might be like in the next century. It is, however, reasonable to speculate on the external factors that may directly influence management decisions in the future. The purpose of this chapter is to formulate management principles needed to guide the development of soil management systems that will sustain and improve the soil resource base in the response to whatever external factors operate in the next century.

A Management Principle for the Twenty-First Century

The agricultural paradigm throughout most of the 20th century has been to produce more food per unit area. In more recent years, specifically the 1980s, the focus has shifted to the most profit per unit area, although, in essence, this is often highly correlated with increased production. Agriculture has not focused well on the energy and material flows within agricultural production systems. The oil crisis of the 1970s forced agriculture to look at energy flows, but the urgency of the 1970s dissipated in the 1980s. Environmental quality issues, particularly ground and surface water quality, has forced agriculture to examine material flows, particularly nutrient flows. This is, to a large extent, the reason for the current interest in sustainable agriculture.

Soil management practices in the 21st century must be formulated on an understanding of the ecosystem concept. Agricultural production systems are managed ecosystems. Although different, they function in accordance with the same set of physical and biological laws as all other ecosystems. The ecological paradigm supports the notion that "in order to understand both the potential productivity and frailty of agricultural systems, we must be aware of the fundamental ecological relationships and processes that are common to all ecosystems" (*2*).

Ecosystems are dynamic systems with characteristic patterns of energy flow, nutrient cycling, and structural change, the structure being defined in terms of species composition, biomass, and spatial organization (*2*). Agricultural ecosystems, or agroecosystems, differ from natural ecosystems in both structure and function in a number of basic ways (*2*):

1. Agroecosystems are highly productive and have huge energy and material subsidies (human and animal labor, fossil fuels, pesticides, chemical fertilizers, and irrigation).

2. Agroecosystems tend to be very simple ecosystems and resemble early stages of succession; they are, therefore, inherently unstable. Stability in an ecosystem results from an increase in diversity and complexity as succession proceeds and the system evolves toward the most stable state possible within the constraints of the environment (called the climax state).

3. Natural ecosystems are self-regulating and self-perpetuating; they display considerable continuity, while agroecosystems do not.

4. Agroecosystems are not adaptive because they consist of unnatural assemblages of species that have not been integrated into a steady-state system by the process of co-evolution.

5. Agroecosystems are not as diverse as natural systems because they do not have a wide range of genetic variability necessary to reduce the impact of a variable environment. Growth and development of crop plants in a field is highly synchronized and requires material and energy subsidies.

6. Agroecosystems are managed by a strategy that varies considerably from nature's balancing of productivity and the resulting progress toward stability.

The application of ecological principles to agricultural production systems presents a plausible means of addressing the sustainability issues facing agriculture now and in the 21st century. Agriculturalists are coming to the realization that a shift toward basic ecological principles is required, but they are insecure with such a movement, in part because the rigid ecological paradigm and a basic lack of understanding of ecological principles. Ecologists insist that new systems of production must be developed that maximize the role of ecological mechanisms and emphasize the use of renewable resources. Agriculturalists are responding to constraints and demands in managed systems often not considered by ecologists in natural ecosystems.

The sustainable agriculture paradigm for the 21st century must be formulated by examining the strengths and weaknesses of managed and natural ecosystems. The exact nature of the sustainable agriculture paradigm is uncertain and will continually evolve. A key to the evolution toward sustainability is to alter the energy and material flows within current agricultural production systems to flows that are based on ecological principles. The major objectives outlined in figure 1 are achievable through judicious soil management options. Prioritizing the issues depicted in the figure, we propose that the management principle for the 21st century is to manage soils in space and time. Within this framework, we offer three specific management principles: (1) farming by soilscapes (soils and landscapes), (2) managing zones within the field, and (3) managing the noncrop period.

Farming by Soilscapes

The principle of farming by soil is to match management practices to specific soil and landscape characteristics and to broaden the agroecosystem to include all control options needed

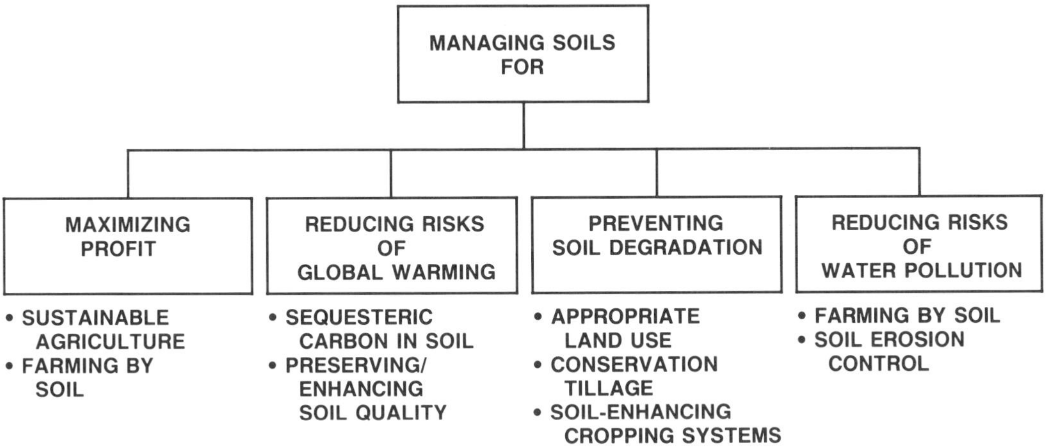

Figure 1. Principle objectives of judicious soil management and technological options to achieve them.

in the system. The first aspect of this principle deals with the fact that farmers farm fields that contain a number of soils and landforms with different levels of inherent productivity and susceptibility to undesirable material flows (erosion, leaching, gaseous losses). Larson and Robert (*6*) give a detailed illustration of the farming-by-soil principle.

Farmers often apply management practices designed for the better soils, resulting, for example, in inappropriate levels of chemical inputs or inadequate erosion control. Yet spatial variability of soils within fields suggests that soil management in the 21st century must include capabilities to adjust management on-the-go using computerized soil maps, aerial photos, and grid maps detailing yield potentials, soil limitations, and pest histories. Equipment will be available to apply pest management technologies with predetermined prescription accuracy as a machine passes over the field. Tillage implements and planters will be adjusted from the tractor cab to allow for prescription tillage and planting adjusted for each soil and landform in the field. When management practices are applied on a soil-by-soil basis, the end result will be improved efficiencies in energy and material flows. This, in turn, will lead to increased profitability for farmers and to a reduction in soil and environmental deterioration.

The second aspect of farming by soilscapes deals with the notion that the controls to a problem occurring within an agricultural field are often located outside the boundaries of the field. Soil erosion control, for example, often involves regional solutions on the scale of a watershed whose boundaries transcend field, farm, or other political boundaries. In another example, it is rare for a farmer to manage field edges, fence rows, woodlots, etc., as a habitat for biological control of pests within a crop field. Soils and landscapes will necessarily be managed in the 21st century across field and political boundaries in an attempt to manage energy and material flows from adjacent ecosystems or ecotypes.

Managing Zones within the Field

In general, agricultural production systems are simplistic. The entire field in the agroecosystem usually is managed as a unit. This management approach defies the notion of management in space and time critical to agricultural sustainability. To manage soils in space and time in order to improve efficiencies of energy and material flows will require farmers to develop management zones within a field. Two very good examples of this principle are the row-interrow management concept developed by Larson (*5*) and alley cropping.

The idea of row-interrow management is to manage the row area differently than the area between the rows. The row area is managed to provide a good soil structure for plant germination, emergence, and root growth as well as appropriate temperature, moisture, fertility, and weed control for the growth and development of the crop. The interrow area is managed to maximize infiltration of water, erosion control, and trafficability and to minimize pests. The interrow area is managed to create a rough surface to maximize infiltration and inhibit germination of weed seeds; be firm for wheel traffic; be seeded to a cover crop to fix nitrogen, suppress weeds, manage beneficial pests, and catch residual nitrates; and be kept low in fertility to deter leaching and discourage weed growth. Examples of this concept include band applications of herbicides, placement of fertilizers rather than broadcast applications, intercropping companion, cover or trap crops in the interrow area, and confining wheel traffic to selected interrow areas. Ridge tillage is a soil management system that exemplifies the row-interrow concept.

Alley cropping is a system that combines the growing of ligneous species with field crops arranged spatially and temporally to provide food and fiber while preserving the soil resource base and minimizing environmental degradation. In Africa, alley cropping involves rotations of food or arable crops grown between hedgerows of preferably leguminous trees or shrubs that are periodically pruned to supply both mulch and fuelwood and reduce shading on the crop (*4, 7*). Interest in agroforestry systems, such as alley cropping, in temperate zones, such as the United States, focuses on the potential to properly manage marginal land and at the same time increase the production of high-value hardwoods (*3*).

The concept of managing zones within the field is not new. The U.S. Department of Agriculture's Soil Conservation Service has used this concept in designing water and erosion control measures since the 1930s (grassed waterways, diversions, filter strips, stripcropping, windbreaks, etc.). However, many of these measures have been under-utilized largely because of the increases in the size of fields and equipment used in today's mechanized agriculture. Management systems such as ridge tillage, zone tillage, intercropping, and alley cropping demonstrate the utility of this principal in agricultural systems.

Managing the Noncrop Period

In many regions of the world, soils are managed intensively during a certain period of the year most conducive to the production of a particular crop. Outside that period, little is done to manage the soil. The problem is that major degradation processes, such as erosion, leaching, and compaction, are often most intense during the noncrop period. In the Corn Belt region of the United States, the noncrop period is October to May. Consider the data reported by Chichester (*1*) that clearly show that the majority of percolation and nitrate losses during the 1972-1975 period occurred in the Keene silt loam soil during the noncrop period.

A key to sustainability in the 21st century will be the extent to which agriculture will develop cultural practices to manage the noncrop period to minimize undesirable material flows from the agroecosystem. To our knowledge, there are currently no efforts directed specifically at developing management systems for the noncrop period. The operational ecological principle is to use the noncrop period to increase the diversity and complexity of the agroecosystem and thereby increase its stability. For example, cover crops could be seeded during the life cycle of a crop to serve ecologically important functions, including erosion control, suppression of pests, alteration of pest cycles, and fixation and biocycling of nutrients. Cover crops would increase the net primary productivity, enhance the detritus food chain, and enhance the organic portion of nutrient cycles. Cover crops could be selected or bred for compatibility with the succeeding crop to serve as a host for organisms important for biological control of pests.

Summary

Agricultural production systems are managed ecosystems that differ considerably from natural ecosystems in structure and function.

As an ecosystem, however, agroecosystems function in accordance with the same set of physical and biological laws as all other ecosystems. Therefore, incorporation of ecological principles from natural ecosystems into agroecosystems is key to development of management systems for the 21st century. Ecology rests on principles, and principles can be applied (Tiedje, personal communication).

There are many examples of ecologically sound concepts and cultural practices that could be implemented in production agriculture for the 21st century. We have proposed three management principles for the 21st century under the general theme of managing soils and landscapes in space and time: farming by soilscapes, managing zones within the field, and managing the noncrop period. New technologies will allow for managing soils in space and time on the basis of soils and their properties, management zones within the field, and processes active during the noncrop period. Such things as tillage, crop rotations, cover crops, alternatives to chemical pest control, grazing management, manipulating genetic variability, and alternative production systems, such as alley cropping, all have ecological strengths and weaknesses that must be sorted out. Successful soil management systems in the 21st century will be those that best incorporate ecological principles, such as those presented here.

REFERENCES CITED

1. Chichester, F. W. 1977. *Effects of increased fertilizer rates on nitrogen content of runoff and percolate from monolith lysimeters.* Journal of Environmental Quality 6: 211-217.
2. Cox, G. W., and M. D. Atkins. 1979. *Agricultural ecology: An analysis of world food production systems.* W. H. Freeman and Company, San Francisco, California. 721 pp.
3. Gold, M. A., and J. W. Hanover. 1987. *Agroforestry in temperate regions.* Agroforestry Systems 5: 109-121.
4. Kang, B. T., A.C.B.M. van der Kruijs, and D. C. Couper. 1989. *Alley cropping for food and crop production in the humid and subhumid tropics.* In B. T. Kang and L. Reynolds [editors] *Alley Farming in the Humid and Subhumid Tropics, Proceedings of an International Workshop.* International Development Research Center, Ottawa, Ontario. pp. 16-26.
5. Larson, W. E. 1964. *Soil parameters for evaluating tillage needs and operations.* Soil Science Society of America Proceedings 38: 118-122.
6. Larson, W. E., and P. Robert. 1990. *Farming by soil.* In R. Lal and F. J. Pierce [editors] *Soil Management for Sustainability.* Soil and Water Conservation Society, Ankeny, Iowa.
7. Okigbo, B. N. 1990. *Sustainable agricultural systems in tropical Africa.* In C. A. Edwards, R. Lal, P. Madden, R. H. Miller, and G. House [editors] *Sustainable Agricultural Systems.* Soil and Water Conservation Society, Ankeny, Iowa. pp. 323-352.

Contributors

R. R. Allmaras
Agricultural Research Service
St. Paul, Minnesota

Charles M. Benbrook
National Research Council
Washington, D.C.

C. F. Bentley
Department of Soil Science (retired)
University of Alberta
Edmonton

A. C. Chang
Department of Soil and Environmental Science
University of California
Riverside

R. M. Cruse
Department of Agronomy
Iowa State University
Ames

R. H. Dowdy
Agricultural Research Service
St. Paul, Minnesota

W. M. Edwards
Agricultural Research Service
Coshocton, Ohio

A. V. Elliot
Department of Agricultural Engineering
Ohio State University
Columbus

W. J. Elliot
Department of Agricultural Engineering
Ohio State University
Columbus

S. A. El-Swaify
Department of Agronomy and Soil Science
University of Hawaii at Monoa
Honolulu

G. R. Foster
Department of Agricultural Engineering
University of Minnesota
St. Paul

P. H. Groenevelt
Department of Land Resource Science
University of Guelph
Guelph, Ontario

S. C. Gupta
Soil Science Department
University of Minnesota
St. Paul

G. H. Heichel
Department of Agronomy
University of Illinois
Urbana

R. Lal
Department of Agronomy
Ohio State University
Columbus

G. W. Langdale
Agricultural Research Service
Watkinsville, Georgia

W. E. Larson
Soil Science Department (retired)
University of Minnesota
St. Paul

W. C. Moldenhauer
World Association of Soil and Water Conservation
Volga, South Dakota

A. L. Page
Department of Soil and Environmental Science
University of California
Riverside

F. J. Pierce
Department of Crop and Soil Sciences
Michigan State University
East Lansing

David Pimentel
Department of Entomology
Cornell University
Ithaca, New York

P. C. Robert
Soil Science Department
University of Minnesota
St. Paul

B. A. Stewart
Agricultural Research Service
Bushland, Texas

P. W. Unger
Agricultural Research Service
Bushland, Texas

D. M. Van Doren
Ohio Agricultural Research and Development Center
Ohio State University
Wooster

Index